Lectures and Essays

VOLUME 1

WILLIAM KINGDON CLIFFORD
EDITED BY LESLIE STEPHEN
AND FREDERICK POLLOCK

CAMBRIDGE
UNIVERSITY PRESS

CAMBRIDGE UNIVERSITY PRESS

Cambridge, New York, Melbourne, Madrid, Cape Town,
Singapore, São Paolo, Delhi, Tokyo, Mexico City

Published in the United States of America by Cambridge University Press, New York

www.cambridge.org
Information on this title: www.cambridge.org/9781108040945

© in this compilation Cambridge University Press 2011

This edition first published 1879
This digitally printed version 2011

ISBN 978-1-108-04094-5 Paperback

LECTURES AND ESSAYS

VOL. I.

C. H. Jeens

Engraved by C.H. Jeens from a Photograph by Barraud & Jerrard.

London Published by Macmillan & Co.

LECTURES AND ESSAYS

BY THE LATE

WILLIAM KINGDON CLIFFORD, F.R.S.

LATE PROFESSOR OF APPLIED MATHEMATICS AND MECHANICS IN UNIVERSITY COLLEGE, LONDON
AND SOMETIME FELLOW OF TRINITY COLLEGE, CAMBRIDGE

EDITED BY

LESLIE STEPHEN AND FREDERICK POLLOCK

WITH an INTRODUCTION by F. POLLOCK

'*La vérité est toute pour tous*'—PAUL-LOUIS COURIER

IN TWO VOLUMES

VOL. I.

London

MACMILLAN AND CO.

1879

CONTENTS

OF

THE FIRST VOLUME.

INTRODUCTION.

PART   PAGE

I. BIOGRAPHICAL 1

II. SELECTIONS FROM LETTERS, ETC. 44

III. BIBLIOGRAPHICAL 67

LECTURES AND ESSAYS.

ON SOME OF THE CONDITIONS OF MENTAL DEVELOPMENT . 75

ON THEORIES OF THE PHYSICAL FORCES 109

ON THE AIMS AND INSTRUMENTS OF SCIENTIFIC THOUGHT . 124

ATOMS 158

THE FIRST AND THE LAST CATASTROPHE 191

THE UNSEEN UNIVERSE 228

THE PHILOSOPHY OF THE PURE SCIENCES 254

INTRODUCTION.

PART I.

BIOGRAPHICAL.

IT is an open secret to the few who know it, but a mystery and a stumbling-block to the many, that Science and Poetry are own sisters; insomuch that in those branches of scientific inquiry which are most abstract, most formal, and most remote from the grasp of the ordinary sensible imagination, a higher power of imagination akin to the creative insight of the poet is most needed and most fruitful of lasting work. This living and constructive energy projects itself out into the world at the same time that it assimilates the surrounding world to itself. When it is joined with quick perception and delicate sympathies, it can work the miracle of piercing the barrier that separates one mind from another, and becomes a personal charm. It can be known only in its operation, and is by its very nature incommunicable and indescribable. Yet this faculty, when a man is gifted with it, seems to gather up the best of his life, so that the man always transcends every work shapen and sent forth by him; his presence is full of it, and it lightens the air his friends breathe; it commands not verbal assent to propositions or intellectual acquiescence in arguments, but the conviction of being in the sphere of a vital force for which nature must make room. Therefore when, being happy in that we knew and saw these things, and have received the imperishable gifts, we must unhappily speak of the friend who gave them as having passed from us, it becomes nothing less than a duty to attempt the impossible task ; to describe that which admits of no description, and communicate that for which words are but blundering messengers. And perhaps it may not be in vain ; for a voice which is in itself weak may strengthen the kindred notes that vibrate in other memories touched by the same power, and those we know to be very many. For this power, when it works for fellowship and not ambition, wins for its wearer the love of all sorts and conditions of men, and this was marked in Clifford by all who had to do with him even a little. More than this, our words may peradventure strike farther, though by no force or skill of their own, and stir some new accord in imaginations favourably attuned for the impulse. The discourses and writings collected in this book will indeed testify to the intellectual grasp and acuteness that went to the making of them. Clifford's earnestness and simplicity, too, are fairly enough presented to the reader, and the clearness of his expression is such that any comment by way of mere explanation would be impertinent. But of the winning felicity of his manner, the varied and flexible play of his thought, the almost boundless range of his human interests and sympathies, his writing tells—at least, so it seems to those who really knew him—nothing or very little. To say a word or two in remembrance of one's friend is but natural; and in these days excuse is hardly needed for saying it in public. But here this is the least part of the matter in hand. Personal desires and aims are merged in the higher responsibility of telling the world that it has lost a man of genius; a responsibility which must be accepted even with the knowledge that it cannot be adequately discharged.

Not many weeks had passed of my first year at Trinity when it began to be noised about that among the new minor scholars there was a young man of extraordinary mathematical powers, and eccentric in appearance, habits, and opinions. He was reputed, and at the time with truth, an ardent High Churchman. I think it was then a more remarkable thing at Cambridge than it would be now, the evangelical tradition of Simeon and his school being still prevalent. This was the first I heard of Clifford; and for some two years he continued to be nothing more to me than a name and a somewhat enigmatic person. In the course of our third year circumstances brought us together: it is difficult to remember the beginnings of a friendship that seems as if it must always have been, but to the best of my recollection there was nothing very sudden or rapid in our closer approach. I should assign about six months as the interval filled by the transition from acquaintance to intimacy. At an early stage in my knowledge of him I remember being struck by the daring versatility of his talk. Even then there was no subject on which he was not ready with something in point, generally of an unexpected kind; and his unsurpassed power of mathematical exposition was already longing to find exercise. I shall be pardoned for giving a concrete instance which may be in itself trivial. In the analytical treatment of statics there occurs a proposition called Ivory's Theorem concerning the attractions of an ellipsoid. The text-books demonstrate it by a formidable apparatus of co-ordinates and integrals, such as we were wont to call a *grind*. On a certain day in the Long Vacation of 1866, which Clifford and I spent at Cambridge, I was not a little exercised by the theorem in question, as I suppose many students have been before and since. The chain of symbolic proof seemed artificial and dead ; it compelled the understanding but failed to satisfy the reason. After reading and learning the proposition one still failed to see what it was all about. Being out for a walk with Clifford, I opened my perplexities to him ; I think I can recall the very spot. What he said I do not remember in detail, which is not surprising, as I have had no occasion to remember anything about Ivory's Theorem these twelve years. But I know that as he spoke he appeared not to be working out a question, but simply telling what he saw. Without any diagram or symbolic aid he described the geometrical conditions on which the solution depended, and they seemed to stand out visibly in space. There were no longer consequences to be deduced, but real and evident facts which only required to be seen. And this one instance, fixed in my memory as the first that came to my knowledge, represents both Clifford's theory of what teaching ought to be, and his constant way of carrying it out in his discourses and conversation on mathematical and scientific subjects. So whole and complete was the vision that for the time the only strange thing was that anybody should fail to see it in the same way. When one endeavoured to call it up again, and not till then, it became clear that the magic of genius had been at work, and that the common sight had been raised to that higher perception by the power which makes and transforms ideas, the conquering and masterful quality of the human mind which Goethe called in one word *das Dämonische.*

A soul eager for new mastery and ever looking forward cares little to dwell upon the past; and Clifford was not much apt to speak of his own earlier life, or indeed of himself at all. Hence I am indebted to his wife and to other friends for what little I am able to say of the time before I knew him. William Kingdon Clifford was born at Exeter on May 4, 1845; his father was·a well-known and active citizen, and filled the office of justice of the peace. His mother he lost early in life; he inherited from her probably some of his genius, and almost certainly the deep-seated constitutional weakness, ill paired with restless activity of nerve and brain, which was the cause of his premature loss. He was educated at Exeter till 1860, when he was sent to King's College, London, not without distinction already won in the University Local Examinations. At school he showed little taste for the ordinary games, but made himself proficient in gymnastics; a pursuit which at Cambridge he carried out, in fellowship with a few like-minded companions, not only into the performance of the most difficult feats habitual to the gymnasium, but into the invention of other new and adventurous ones. But (as he once said himself of Dr. Whewell) his nature was to touch nothing without leaving some stamp of invention upon it. His accomplishments of this kind were the only ones in which he ever manifested pride. When he took his degree there was a paragraph in 'Bell's Life' pointing out, for the rebuke of those who might suppose manly exercises incompatible with intellectual distinction, that the Second Wrangler, Mr. Clifford, was also one of the most daring athletes of the University. This paragraph gave him far more lively pleasure than any of the more serious and academical marks of approval which he had earned. In 1869 he wrote from Cambridge :—' I am at present in a very heaven of joy because my corkscrew was encored last night at the assault of arms : it consists in running at a fixed upright pole which you seize with both hands and spin round and round descending in a corkscrew fashion.' In after years he did not keep up his gymnastic practice with anything like regularity ; but he was with great difficulty induced to accept the necessity of completely abandoning it when it was known to be positively injurious to his health. A friend who was his companion in gymnastics writes to me :—' His neatness and dexterity were unusually great, but the most remarkable thing was his great strength as compared with his weight, as shown in some exercises. At one time he could pull up on the bar with either hand, which is well known to be one of the greatest feats of strength. His nerve at dangerous heights was extraordinary. I am appalled now to think that he climbed up and sat on the cross bars of the weathercock on a church tower, and when by way of doing something worse I went up and hung by my toes to the bars he did the same.'

At King's College Clifford's peculiar mathematical

abilities came to the front, but not so as to exclude attention to other subjects. He was at various times and in various ways marked out for honourable mention in classics, modern history, and English literature. His knowledge of the classics, though he did not cultivate the niceties of scholarship, was certainly as sound and extensive as that of many professedly classical students ; and, like all his knowledge, it was vital. If he made use of it for quotation or otherwise, it was not because the passage or circumstance was classical, but because it was the thing he wanted to illustrate his own thought. Of history he knew a good deal; he was fond of historical reading throughout his life, and had a ready command of parallels and analogies between widely remote times and countries, sometimes too ingenious to bear criticism. I doubt if he studied historical works critically ; it seems to me that he regarded history in a poetical rather than a scientific spirit, seeing events in a series of vivid pictures which had the force of present realities as each came in turn before the mind's eye. Thus he threw himself into the past with a dramatic interest, and looked on the civilized world as a field where the destinies of man are fought out in a secular contest between the powers of good and evil, rather than as a scene of the development and interaction of infinite and infinitely complex motives. This indeed, in a meagre and far cruder form, is essentially the popular view ; the sort of history upon which most people are still brought up divides men, actions, and institutions into good and bad according to the writer's present notions of what might and ought to be, and distributes blessing and cursing without more

ado. Only Clifford, accepting to some extent the
popular or pictorial way of looking at history, took on
most questions the unpopular side, and so found himself in collision with current opinions. He had a fair
general knowledge of English literature (by which I
mean considerably more than is yet supposed necessary
for an Englishman's education), with a preference for
modern poetry, and especially for such as gave expression to his own ideas. Milton's prose had also a
special attraction for him. I do not think he cared
much for the use of language as a fine art, though he
had a great appreciation of arrangement and composition. His own style, always admirably clear and often
eloquent, was never elaborate ; for we cannot fairly
count the studied ornament of his College declamations,
which were not only produced while he was an undergraduate, but for an occasion which justified some
special aiming at rhetorical effect. Much of his best
work was actually spoken before it was written. He
gave most of his public lectures with no visible prepa-
ration beyond very short notes, and the outline seemed
to be filled in without effort or hesitation. Afterwards
he would revise the lecture from a shorthand-writer's
report, or sometimes write down from memory almost
exactly what he had said. It fell out now and then,
however, that neither of these things was done ; and in
such cases there is now no record of the lecture at all.
Once or twice he tried writing part of the lecture
beforehand, but found it only an embarrassment in the
delivery. I believe the only one wholly put in writing
in the first instance was *Ethics of Religion*, which he

was unable to deliver himself. I cannot find anything showing early aptitude for acquiring languages ; but that he had it and was fond of exercising it in later life is certain. One practical reason for it was the desire of being able to read mathematical papers in foreign journals; but this would not account for his taking up Spanish, of which he acquired a competent knowledge in the course of a tour to the Pyrenees. When he was at Algiers in 1876 he began Arabic, and made progress enough to follow in a general way a course of lessons given in that language. He read modern Greek fluently, and at one time he was curious about Sanskrit. He even spent some time on hieroglyphics. A new language is a riddle before it is conquered, a power in the hand afterwards: to Clifford every riddle was a challenge, and every chance of new power a divine opportunity to be seized. Hence he was likewise interested in the various modes of conveying and expressing language invented for special purposes, such as the Morse alphabet and shorthand. One of his ideas about education was that children might learn these things at an early age, perhaps in play, so as to grow up no less familiar with them than with common printing and writing. I have forgotten to mention his command of French and German, the former of which he knew very well, and the latter quite sufficiently ; I think his German reading was mostly in the direction of philosophy and mathematics.

In 1863 Clifford came up with a minor scholarship to Trinity College, Cambridge ; in his third year (to continue for the present on the line of his literary accomplishments) he won the College declamation prize[1] with a very brilliant discourse on Sir W. Raleigh, partly cast in the form of quasi-dramatic dialogues, and accordingly had to deliver the annual oration at the Commemoration of Benefactors in December. His subject was a panegyric of the late Master of the College, Dr. Whewell, whose death was then recent. It was treated in an original and unexpected manner, Dr. Whewell's claim to admiration and emulation being put on the ground of his intellectual life exemplifying in an eminent degree the active and creating faculty. 'Thought is powerless except it make something outside of itself: the thought which conquers the world is not contemplative but active. And it is this that I am asking you to worship to-day.' Taking this oration as a whole, it must be considered as a *tour de force*, giving glimpses and undetermined promises of speculative power. But there occurred in it an apologue which caught the attention of some good judges at the time, and so well illustrates the fanciful and sportive side of Clifford's mind that I shall here transcribe it.

'Once upon a time—much longer than six thousand years ago—the Trilobites were the only people that had eyes; and they were only just beginning to have them, and some even of the Trilobites had as yet no signs of coming sight. So that the utmost they could know was that they were living in darkness, and that perhaps there was such a thing as light. But at last one of them got so far advanced that when he happened to

---

[1] He was bracketed with Mr. C. A. Elliott for the first prize: but (I now forget for what reason) the office of delivering the Oration fell to Clifford alone.

come to the top of the water in the daytime he saw the sun. So he went down and told the others that in general the world was light, but there was one great light which caused it all. Then they killed him for disturbing the commonwealth ; but they considered it impious to doubt that in general the world was light, and that there was one great light which caused it all. And they had great disputes about the manner in which they had come to know this. Afterwards another of them got so far advanced that when he happened to come to the top of the water in the night-time he saw the stars. So he went down and told the others that in general the world was dark, but that nevertheless there was a great number of little lights in it. Then they killed him for maintaining false doctrines : but from that time there was a division amongst them, and all the Trilobites were split into two parties, some maintaining one thing and some the other, until such time as so many of them had learned to see that there could be no doubt about the matter.'

The interpretation was barely indicated on this occasion ; but it is worked out in another Cambridge MS. which must have been written shortly afterwards, and in which the apologue stands first as a kind of text. It was nothing less than a theory of the intellectual growth of mankind ; and the position was that, as the physical senses have been gradually developed out of confused and uncertain impressions, so a set of intellectual senses or *insights* are still in course of development, the operation of which may ultimately be expected to be as certain and immediate as our ordinary sense-perceptions.

This theory may be traced in the discourse ' On some of the Conditions of Mental Development,' delivered in March, 1868, which stands first in the present collection; and for that reason I make special mention of it. Otherwise it was only one inventive experiment among many. I should far exceed my limits if I were to attempt any account of the various forms of speculation, physical, metaphysical, social, and ethical, through which Clifford ranged in the first few years after his degree. Not that he was constantly changing his opinions, as a superficial observer might have thought; he was seeking for definite principles, and of set purpose made his search various and wide-spread. He had a singular power of taking up any theory that seemed to him at all worth investigating, realizing it, working it out, and making it completely his own for the time being, and yet all the while consciously holding it as an experiment, and being perfectly ready to give it up when found wanting.

Clifford's mathematical course at Cambridge was a struggle between the exigencies of the Tripos and his native bent for independent reading and research going far beyond the subjects of the examination; and the Tripos had very much the worst of it. If there was any faculty in which he was entirely wanting, it was the examination-faculty. On this subject I am not competent to speak with certainty, but it is my belief that, from the point of view to which the class-list is an end in itself, Clifford omitted most of the things he ought to have read, and read everything he ought not to have read. Nevertheless his powers of original work carried him so far that he came out Second Wrangler in the

Tripos of 1867, and was also Second Smith's Prizeman. I am fortunately able to quote on this head the statement of one of our first living analysts, Professor Sylvester:—

'Like the late Dr. Whewell, Professor Clerk Maxwell, and Sir William Thomson, Mr. Clifford was Second Wrangler at the University of Cambridge. I believe there is little doubt that he might easily have been first of his year had he chosen to devote himself exclusively to the University curriculum instead of pursuing his studies, while still an undergraduate, in a more extended field, and with a view rather to self-culture than to the acquisition of immediate honour or emolument.'

This pursuit of knowledge for its own sake, and without even such regard to collateral interests as most people would think a matter of common prudence, was the leading character of Clifford's work throughout his life. The discovery of truth was for him an end in itself, and the proclamation of it, or of whatever seemed to lead to it, a duty of primary and paramount obligation. This had something to do with the fascination of his teaching; he never seemed to be imposing dogmas on his hearers, but to be leading them into the enjoyment of a common possession. He did not tell them that knowledge was priceless and truth beautiful; he made them feel it. He gave them not formulas, but ideas. Again I can appeal to a witness of undoubted authority. The following words were written in 1871 by a man in no way given to unmeasured expression of his mind, and as eminent in mathematical physics as the author of the statement I have already cited is in pure mathematics, I mean Professor Clerk Maxwell:—

'The peculiarity of Mr. Clifford's researches, which in my opinion points him out as the right man for a chair of mathematical science, is that they tend not to the elaboration of abstruse theorems by ingenious calculations, but to the elucidation of scientific ideas by the concentration upon them of clear and steady thought. The pupils of such a teacher not only obtain clearer views of the subjects taught, but are encouraged to cultivate in themselves that power of thought which is so liable to be neglected amidst the appliances of education.'

I shall not attempt to enter in more detail on the amount and character of Clifford's subsequent contributions to mathematical science, having reason to hope that this task will shortly be undertaken by competent hands and in a more appropriate connexion. But in an introduction to his philosophical writings it is fitting to call attention to the manner in which he brought mathematical conceptions to bear upon philosophy. He took much pleasure in the speculative constructions of imaginary or non-Euclidean systems of space-relations which have been achieved by Continental geometers, partly because they afforded a congenial field for the combined exercise of scientific intuition and unbridled fancy. He liked talking about imaginary geometry, as a matter of pure amusement, to anyone interested in it. But at the same time he attached a serious import to it. He was the first in this country, as Helmholtz in Germany, to call attention to the philosophical importance of these new ideas with regard to the question of the nature and origin of geometrical knowledge. His opinion on this point is briefly expressed in the lectures *On the Philosophy of the Pure*

Sciences. He intended to recast and expand these, and doubtless would have amplified this particular discussion. It will be seen that he considered Kant's position in the matter of 'transcendental æsthetic' to be wholly unassailable if it was once admitted that geometrical knowledge is really exact and universal. The ordinary arguments for the derivative nature of axioms appeared to him ingenious but hopeless attempts to escape from this fatal admission. And it may be said in general terms that he had a much fuller appreciation of the merit and the necessity of Kant's work than most adherents of the English school of psychology. Of course I do not include Professor Huxley, whose testimony to Kant in his little book on Hume is as unmistakable as it is weighty.

Few words will suffice to set down the remaining facts of Clifford's life, or what we are accustomed to call facts because they can be dated and made equally known to everybody, as if that made them somehow more real than the passages and events which in truth decide the issues of life and fix the courses of a man's work. In 1868 he was elected a Fellow of Trinity College, and after spending rather more than two years at Cambridge, he was in 1871 appointed to the Professorship of Applied Mathematics at University College, London. Meanwhile he had taken part in the English Eclipse expedition of 1870 : his letters of that time show keen enjoyment of the new experience of men and cities, and of the natural beauty of the Mediterranean coasts, which he was to visit again, as fate would have it, only on the sad and fruitless errand of attempting to recover strength when it was too late. In June, 1874, he was elected a Fellow of the Royal Society ; he might have been proposed at a much earlier time, but had then declined, turning it off with the remark that he did not want to be respectable yet. And such was the absence in him of anything like vanity or self-assertion, that when his scruples were overcome, and his election took place, he was the last person from whom his friends heard of it. I did not know it myself till several months later. On April 7, 1875, he married Lucy, daughter of Mr. John Lane, and granddaughter of Brandford Lane, of Barbados. This was the occasion of the only voluntary leave of absence he ever took from his lectures at University College, when he characteristically informed his class that he was obliged to be absent on important business which would probably not occur again. Clifford's house was thenceforward (as, indeed, his rooms, both at Cambridge and in London, had already been) the meeting-point of a numerous body of friends, in which almost every possible variety of taste and opinion was represented, and many of whom had nothing else in common. The scientific element had naturally a certain predominance; and with Clifford, as with other men, a close friendship implied, as a rule, some sort of general coincidence in sentiments and aims, personal and intellectual concord being apt to go together. But he cared for sympathy, not for agreement; coincidence in actual results was indifferent to him. He wrote of a very near and dear friend (G. Crotch, of St. John's College, Cambridge), whose death preceded his own by some years: 'We never agreed upon results, but we always used the same method with the same object.' Much more would it be an utter mistake to suppose that Clifford was a scientific fanatic who reserved his social qualities for such persons as happened to accept his theories, or that he could not be at his ease and make the charm of his presence felt among those who did not care for theories at all. It was possible to take offence at certain passages in his writings, but impossible not to like the man; and some of those to whom Clifford's published opinions were naturally most repugnant, but who had the opportunity of personal intercourse with him, were by no means the last to express their sympathy and anxiety when the threatenings of the disease which carried him off became apparent. This charm remained with him to his very last days; even when he was in an enfeebled and almost prostrate condition there were those who conceived for him and his, upon sudden and casual acquaintance, an affection and goodwill which bore such fruit of kindly deeds as men usually look for only from the devotion ripened by long familiarity. Something of this was due to the extreme openness and candour of his conversation; something to the quickness with which he read the feelings of others, and the delicacy and gentleness with which he adapted himself to them; something, perhaps most, to a certain undefinable simplicity in which the whole man seemed to be revealed, and the whole moral beauty of his character to be grounded. It was by this simplicity, one may suppose, that he was endeared from his early days to children. He always took delight in being with them, and appeared to have a special gift of holding their attention. That he did not live to teach his own children is deeply to be regretted not only for their sake, but in the interest of education as a science and an art.

What he could do for the amusement of children (and of
all persons healthy enough not to be ashamed of childish-
ness) was shown to the world in his contributions to a
collection of fairy tales called 'The Little People.' One
of these ('The Giant's Shoes,' reprinted in the second
part of this Introduction) is one of the choicest pieces of
pure nonsense ever put together; and he doubtless en-
joyed writing it as much as any child could enjoy hearing
it. A children's party was one of Clifford's greatest
pleasures. At one such party he kept a wax-work
show, children doing duty for the figures; but he re-
proached himself for several days afterwards because he
had forgotten to wind up the Siamese twins. He seemed
to have an inexhaustible store of merriment at all
times: not merely a keen perception of the ludicrous,
but an ever fresh gaiety and gladness in the common
pleasures of life. His laughter was free and clear like
a child's, and as little restrained by any consideration of
conventional gravity. And he carried his mirth and
humour into all departments of life, by no means
excepting philosophy. When he came home from the
meetings of the Metaphysical Society (attending which
was one of his greatest pleasures, and most reluctantly
given up when going abroad after sunset was forbidden
him), he would repeat the discussion almost at length,
giving not only the matter but the manner of what had
been said by every speaker, and now and then making
his report extremely comic by a touch of plausible
fiction. There was an irresistible affectation of innocence
in his manner of telling an absurd story, as if the drollery
of it were an accident with which he had nothing to do.
It was hardly possible to be depressed in his company:

and this was so not only in his best days, but as long as he had strength to sustain conversation at all. The charm of his countenance and talk banished for the time the anxiety we felt for him (only too justly) whenever we were not with him.

On the intellectual side this character of simplicity manifested itself in the absolute straightforwardness of everything he said and did ; and this, being joined to subtlety and a wide range of vision, became in speculation and discussion a very formidable power. If there was anything for which he had no toleration, and with which he would enter into no compromise, it was insincerity in thought, word, or deed. He expressed his own opinions plainly and strongly because he held it the duty of every man so to do ; he could not discuss great subjects in a half-hearted fashion under a system of mutual conventions. As for considerations of policy or expediency that seemed to interfere in any way with the downright speaking of truth for the truth's sake, he was simply incapable of entertaining them. 'A question of right and wrong,' he once wrote to me, ' knows neither time, place, nor expediency.' Being always frank, he was at times indiscreet ; but consummate discretion has never yet been recognized as a necessary or even a very appropriate element of moral heroism. This must be borne in mind in estimating such passages of his writings as, judged by the ordinary rules of literary etiquette, may seem harsh and violent.

Personal enmity was a thing impossible to Clifford. Once he wrote : 'A great misfortune has fallen upon me ; I shook hands with —— I believe if all the murderers and all the priests and all the liars in the

world were united into one man, and he came suddenly
upon me round a corner and said, " *How do you do ?* "
in a smiling way, I could not be rude to him upon the
instant.' And it was the bare truth. Neither did he
ever make an enemy that I know of ; I do not count
one or two blundering attacks which, however far they
might go beyond the fair bounds of controversy or
satire, were made by people who only guessed at the
man from a superficial inspection of his writings, and
were incapable of understanding either. Yet he carried
about with him as deadly a foe as could have been
wished him by any of those who fear and hate the light
he strove so manfully to spread abroad. This was the
perilous excess in his own frame of nervous energy over
constitutional strength and endurance. He was able to
call upon himself, with a facility which in the result was
fatal, for the expenditure of power in ways and to an
extent which only a strong constitution could have
permanently supported ; and here the constitution was
feeble. He tried experiments on himself when he ought
to have been taking precautions. He thought, I believe,
that he was really training his body to versatility and
disregard of circumstances, and fancied himself to be
making investments when he was in fact living on his
capital. At Cambridge he would constantly sit up most
of the night working or talking. In London it was not
very different, and once or twice he wrote the whole
night through ; and this without any proportionate re-
duction of his occupations in more usual hours. The
paper on ' The Unseen Universe ' was composed in this
way, except a page or two at the beginning, at a single
sitting which lasted from a quarter to ten in the evening

till nine o'clock the following morning. So, too, was the article on Virchow's address. But Clifford's rashness extended much farther than this one particular. He could not be induced, or only with the utmost difficulty, to pay even moderate attention to the cautions and observances which are commonly and aptly described as taking care of one's self. Had he been asked if it was wrong to neglect the conditions of health in one's own person, as well as to approve or tolerate their neglect on a larger scale, he would certainly have answered yes. But to be careful about himself was a thing that never occurred to him. Even when, in the spring of 1876, distinct and grave indications of pulmonary disease were noted, his advisers and friends could hardly persuade him that there was anything more serious than could be set right by two or three weeks' rest in the country. Here, however, there came into play something more than incredulity or indifference; the spirit of the worker and inventor rebelled against thus being baffled. His repugnance was like that of a wounded soldier who thinks himself dishonoured if he quits the field while his limbs can bear him. Reluctantly and almost indignantly he accepted six months' leave of absence, and spent the summer of that year in a journey to Algiers and the south of Spain. He came back recruited for the time, and was allowed to winter in England on pledges of special care and avoidance of exposure. These were in the main observed, and so matters went on for a year and a half more, as it seemed with fair prospects of ultimate recovery and tolerably secure enjoyment of life. What mischief was already done could not be undone; but the spread of it seemed in a way to be permanently arrested. But in the early months of last year there came a sudden change for the worse. His father's death, which happened at this time, was a grievous blow, and the conjunction of this with exciting literary work, done under pressure of time, threw upon him a strain which he was wholly unable to resist. The essay on Virchow's address, which closes the present collection, is both in my opinion and in that of other and more competent judges one of Clifford's best and most mature performances. But it was produced at a fearful cost, we have already seen in what manner. A few days after the MS. had left his hands he received a peremptory warning that he was in a state of such imminent danger that he must give up all work and leave England forthwith. This time the warning was too stern to admit of doubt or even delay. Yet, while the necessary preparations were in hand, he would not leave his official duties until he actually broke down in the attempt to complete a lecture. He was now suffering, not from any inroad of specific local disease, but from a rapid and alarming collapse of general strength which made it seem doubtful if he could live many weeks. But his constitutional frailty was accompanied withal by a wonderful power of rallying from prostration; and one could not help entertaining a dim hope, even to the last, that this vitality was somehow the deepest thing in his nature, and would in the long run win the day. In April, 1878, Clifford and his wife left England for the Mediterranean; the accounts they sent home were various and often anxious; but after voyages and short halts which embraced Gibraltar, Venice, and

Malta, they rested for some weeks at Monte Generoso, and there for the first time there was the appearance of steady improvement setting in. From this place Clifford wrote long letters with his own hand, full of his usual spirit and manifold interest in everything about him. I may mention here that his letters were the more valuable because they were always spontaneous and could very seldom be counted on beforehand. He wrote quickly and easily; and yet for some obscure reason letter-writing, especially as a matter of business, was beyond measure irksome and difficult to him. He would rather take almost any trouble than answer a letter, and the painfulness of answering was at its height when (as pretty often happened) old acquaintances applied to him for testimonials. For in this case it was aggravated by the utter impossibility of lending himself to the petty exaggerations and dissimulations which custom allows to pass current for such purposes, and which are almost thought to be required by civility. One such application, from a man he had known before but had lost sight of, vexed him extremely; he did not know what to do with it, for he could honestly have certified only as to the past, and he carried the letter about with him till it was ragged, being newly vexed every time he saw it. There were many letters of friends which he regretted to the last not having answered. Several received in the last months or weeks of his life he intended to answer if he had ever become strong enough. Yet now and then he would write unsought to some one he was intimate with, and throw himself completely into his letter; and then his descriptions were so full of life and colour that they might

well be taken as models by anyone minded to study
the art of correspondence, not uncommonly alleged to
be lost since the introduction of cheap and rapid com-
munications.   Such letters he sent to England from
Spain and Sicily in 1870, and from Algiers in 1876.
Some of them are printed farther on.

In August, 1878, there being signs of improvement,
and a warm climate not being judged necessary or very
desirable at that season, leave was given for a short
return to England.   Clifford came home looking very
ill and feeble to ordinary observation, but much better
to those who had seen him before he started.   He was
incapable of continuous exertion of any kind, but much
of the old animation had come back, and his conversation had lost nothing of its vigour and brilliancy.
The object of the summer journey had been rest and
freedom from care above all things : now it was planned
that with the first days of autumn he should again go
in search of conditions which might be not only rest-
giving but curative.   But all plans were cut short by a
relapse which took place late in September, induced
by fatigue.   From that day the fight was a losing one,
though fought with such tenacity of life that sometimes
the inevitable end seemed as if it might yet be put far
off.   Clifford's patience, cheerfulness, unselfishness, and
continued interest in his friends and in what was going
on in the world, were unbroken and unabated through
all that heavy time.   Far be it from me, as it was far
from him, to grudge to any man or woman the hope or
comfort that may be found in sincere expectation of a
better life to come.   But let this be set down and remembered, plainly and openly, for the instruction and

rebuke of those who fancy that their dogmas have a monopoly of happiness, and will not face the fact that there are true men, ay and women, to whom the dignity of manhood and the fellowship of this life, undazzled by the magic of any revelation, unholpen of any promises holding out aught as higher or more enduring than the fruition of human love and the fulfilment of human duties, are sufficient to bear the weight of both life and death. Here was a man who utterly dismissed from his thoughts, as being unprofitable or worse, all speculations on a future or unseen world; a man to whom life was holy and precious, a thing not to be despised, but to be used with joyfulness; a soul full of life and light, ever longing for activity, ever counting what was achieved as not worthy to be reckoned in comparison of what was left to do. And this is the witness of his ending, that as never man loved life more, so never man feared death less. He fulfilled well and truly that great saying of Spinoza, often in his mind and on his lips: *Homo liber de nulla re minus quam de morte cogitat.*

One last stand was made, too late to be permanently successful (if ever it could have so far availed), but yet not wholly in vain. At the opening of the present year Clifford's remnant of strength was visibly diminishing. The peril of attempting a journey was great, but no peril could be greater than that which he already lay in. Medicine had no new thing to recommend, and almost nothing to forbid: a last experiment could only be tried. Clifford sailed for Madeira, his friends hardly expecting him to live out the voyage. Of the friendship and devotion that accompanied and tended him there it is not fitting that I should speak. So it was, however, that he arrived safely in the island, and some weeks were added to his life. The change from the bitterest of recent English winters to the fair and temperate air of Madeira had no power to restore the waning forces; but it enabled him to spend his last days in ease and comparative enjoyment. He could once more look on the glories of a bountiful world, and breathe under a free sky. Something of spirit and even of strength revived; his powers of conversation, which had been restrained by mere physical weakness in his last days in England, returned to some extent, and in that short time, with all the disadvantages of a stranger and an invalid, he made new friends: one such (though in spirit not a stranger before) of whose friendship even he might have been proud. There was a glimmer of hope, faint, uncertain, but perceptible; there was a possibility that if amendment once began, it might go farther than we had dared to speculate upon. But it was not to be. In the last days of February we learnt that his condition was hopeless; on the 3rd of March the end came. For a week he had known that it might come at any moment, and looked to it steadfastly. So calmly had he received the warning which conveyed this knowledge that it seemed at the instant as if he did not understand it. He gave careful and exact directions as to the disposal of his works, which are partly carried out in these volumes, and, it is hoped, will be substantially fulfilled as to his mathematical remains also. His work was, indeed, the only thing personal to himself that he took much thought for; and that not because it was his own possession, but because he felt that it was his own to do and to make a possession for others.

He loved it for the work's and the truth's sake, not for his own. More than this, his interest in the outer world, his affection for his friends and his pleasure in their pleasures, did not desert him to the very last. He still followed the course of events, and asked for public news on the morning of his death : so strongly did he hold fast his part in the common weal and in active social life.

It has been mentioned how unwilling Clifford was to throw up, even under necessity, his work at University College. His friends and colleagues there were equally unwilling to lose him ; and when it became evident that he could never permanently resume his lectures, they still cast about for means to retain him as one of their number. In the present year the Senate, in reviewing the whole question of the teaching of mathematics and physics, recommended that Clifford should ' remain in possession of his chair, and that if, against the expectation, but in accordance with the most earnest desire of his colleagues, he should so far recover health as to be able to lecture, he should be invited to lecture upon special subjects in mathematics, to which he could bring his own rare qualities of mind without being subjected to any strain of constant necessary work.' This recommendation only awaited the assent of the Council to take effect, and that assent would almost certainly have been given ; but before the matter could be submitted to the Council it was known that the time of expectation was over, and desire quelled by the final certainty of loss.

The essays here brought together represent, with few if any exceptions, the general view of the world and human knowledge which Clifford had definitely arrived at in his later years. I do not mean that he had got a fixed set of results and meant to rest in them; he admitted no finality of that sort. But he did believe very decidedly that the difference between right and wrong method is everywhere important, and that there is only one right method for all departments of knowledge. He held that metaphysical and theological problems ought to be discussed with exactly the same freedom from preconceived conclusions and fearlessness of consequences as any other problems. And he further held that, as the frank application of the right method of search to the physical sciences has put them on a footing of steady progress, though they differ in the amount and certainty of the knowledge already won in their respective fields, so the like effects might be expected when philosophical speculation was taken in hand by the light of science and with scientific impartiality and earnestness. For the popular or unscientific rhetoric which frequently assumes the name of philosophy Clifford had as much contempt as he permitted himself to feel for anything. Once he said of an acquaintance who was believed to be undertaking something in this kind: 'He is writing a book on metaphysics, and is really cut out for it; the clearness with which he thinks he understands things and his total inability to express what little he knows will make his fortune as a philosopher.' But he never accepted, and I do not think he was ever tempted to accept, the doctrine that all metaphysical inquiries ought to be put aside as unprofitable. Indeed he went beyond most English psychologists, though in a general way he must be classed with the English school, in his estimate of the possibility of constructing a definite metaphysical system on scientific principles. With regard to the application of his philosophical ideas to theological conceptions, it may perhaps be said that he aimed at doing for dogmatic and natural theology something like what the Tübingen school in Germany have done for historical theology, namely bringing them to the light of unbiassed common sense, including therein as an important element the healthy moral sense of civilized men. Whether Clifford had any feeling that his line of work was complementary to the historical criticism of dogmas I cannot say : but so it was that he paid no special attention to the historical side of these questions, either because it did not particularly interest him, or because he thought it outside his competence. In ethics, on the other hand, he attached the utmost importance to the historical facts of moral culture as affording the key of the speculative position and indicating the profitable directions of inquiry. And it may be noted as an instance of the freshness and openness of his mind that the importance of this point of view, set forth in *The Scientific Basis of Morals* and the papers following it, was perceived by him only after he left Cambridge. The main points of the last-named essay were stated by Clifford himself in a letter written when he had nearly finished it. He described it as ' showing that moral maxims are ultimately of the same nature as the maxims of any other craft : if you want to live together successfully, you must do so-and-so. . . . . That conscience is developed out of experience by healthy natural processes. . . . . That responsibility is founded on such order as we can observe, and not upon such disorder as

we can conjecture.' This is quite a different line from
that which his speculations on the nature of duty were
wont to take at Cambridge, both in the conversations I
remember, and in various MS. fragments of that period
which are now before me.

A letter of the autumn of 1874, written by Clifford
to his wife during their engagement, bears upon his
practical conception of ethics and is otherwise interesting. 'At the Savile I found C., who had just done
dinner, but sat down while I ate mine, and we solved
the universe with great delight until A. came in and
wanted to take him off to explain coins to somebody.
Of course I would not let him go. . . . . We walked
about in the New Road solving more universe. He says
the people in the middle ages had a closer connexion
between theory and practice; a fellow would get a
practical idea into his head, be cock-sure it was right,
and then get up and snort and just have it carried
through. Nowadays we don't have prophets with the
same fire and fervour and insight. To which it may be
said that our problems are infinitely more complex, and
that we can't be so cock-sure of the right thing to do.
He quoted the statesmanship of the great emperors,
e.g., Frederic II. ; and some of the saints, as St. Francis
and St. Catherine of Siena. Still there is room for some
earnest person to go and preach around in a simple way
the main straightforward rules that society has unconsciously worked out and that are floating in the air ;
to do as well as possible what one can do best; to work
for the improvement of the social organization ; to seek
earnestly after truth and only to accept provisionally
opinions one has not inquired into ; to regard men as

comrades in work and their freedom as a sacred thing; in fact, to recognize the enormous and fearful difference between truth and falsehood, right and wrong, and how truth and right are to be got at by free inquiry and the love of our comrades for their own sakes and nobody else's. Mazzini has done a great deal in this direction, and formed the conception of the world as a great workshop where we all have to do our best to make something good and beautiful with the help of the others. Such a preaching to the people of the ideas taught by the great Rabbis was (as near as we can make out) the sort of work that Christ did; but he differed from the Rabbis and resembled all other Jew prophets in not being able to stand priests.'

It will not be amiss to go back to the time when we left Clifford celebrating the late Master of Trinity in parables, and to take up more continuously than we have yet done the growth of his philosophic ideas. Before he took his degree, and I think for some little time after, he was (as before mentioned) a High Churchman; but there was an intellectual and speculative activity about his belief which made it impossible that it should remain permanently at that stage. On the one hand he acquired a far more accurate knowledge of Catholic theology than is often met with in England even among those who discuss theological questions; he was pretty well read in S. Thomas Aquinas, and would maintain the Catholic position on most points with extreme ingenuity, not unfrequently adding scientific arguments and analogies of his own. On the other hand, believing from the first in the unity or at least the harmony of all truth, he never slackened in the pursuit of scientific knowledge and ideas. For a while he experimented in schemes for the juxtaposition of science and dogma. Religious beliefs he regarded as outside the region of scientific proof, even when they can be made highly probable by reasoning; for, as he observes in a MS. fragment of this time, they are received and held not as probable but as certain. And he actually defined superstition as 'a belief held on religious or theological grounds, but capable of scientific proof or disproof.' He also held that there was a special theological faculty or insight, analogous to the scientific, poetic, and artistic faculty; and that the persons in whom this genius is exceptionally developed are the founders of new religions and religious orders. He seems to have been always and equally dissatisfied with attempts at proving theological propositions, especially in the usual manner of Protestant divinity, and with the theological version of natural history commonly called Natural Theology. When or how Clifford first came to a clear perception that this position of quasi-scientific Catholicism was untenable I do not exactly know; but I know that the discovery cost him an intellectual and moral struggle, of which traces may be found here and there in his essays. It is not the case, however, that there was any violent reaction or rushing to an opposite extreme. Some time elapsed before his philosophical opinions assumed their final consistency; and in truth what took place was not a reaction, but the fuller development of principles which had been part of his thoughts ever since he began to think for himself.

Meanwhile he was eagerly assimilating the ideas

which had been established as an assured possession of biological science by Mr. Darwin, and the kindred ones already at an earlier time applied and still being applied to the framing of a constructive science of psychology, and to the systematic grouping and gathering together of human knowledge, by Mr. Herbert Spencer ; who had, in Clifford's own words, ' formed the conception of evolution as the subject of general propositions applicable to all natural processes.' Clifford was not content with merely giving his assent to the doctrine of evolution : he seized on it as a living spring of action, a principle to be worked out, practised upon, used to win victories over nature, and to put new vigour into speculation. For two or three years the knot of Cambridge friends of whom Clifford was the leading spirit were carried away by a wave of Darwinian enthusiasm : we seemed to ride triumphant on an ocean of new life and boundless possibilities. Natural Selection was to be the master-key of the universe ; we expected it to solve all riddles and reconcile all contradictions. Among other things it was to give us a new system of ethics, combining the exactness of the utilitarian with the poetical ideals of the transcendentalist. We were not only to believe joyfully in the survival of the fittest, but to take an active and conscious part in making ourselves fitter. At one time Clifford held that it was worth our while to practise variation of set purpose; not only to avoid being the slaves of custom, but to eschew fixed habits of every kind, and to try the greatest possible number of experiments in living to increase the chances of a really valuable one occurring and being selected for preservation. So much of this theory as he ever gave to the world will be found in the discourse *On Some Conditions of Mental Development*; and I do not know that he would ever have deliberately committed himself to anything more than is there propounded. One practical deduction was that education ought to be directed not to mere instruction, but to making people think and act for themselves; and this Clifford held to be of special importance in the case of women, where the cultivation of independent power is too commonly neglected or even purposely discouraged. 'It seems to me,' he once wrote, 'that the thing that is wanting in the education of women is not the acquaintance with any facts, but accurate and scientific habits of thought, and the courage to think that true which appears to be unlikely. And for supplying this want there is a special advantage in geometry, namely that it does not require study of a physically laborious kind, but rather that rapid intuition which women certainly possess; so that it is fit to become a scientific pursuit for them.'

The duty of independence and spontaneous activity conceived by Clifford as being revealed by the philosophy of evolution was reinforced from another side by the reading of Mazzini; and the result was a conception of freedom as the one aim and ideal of man. This freedom was a sort of transfigured blending of all powers of activity and progress; it included republicanism as opposed to the compulsory aspect of government and traditional authority in general, but was otherwise not bound to any particular theory in politics. Indeed it forbade binding oneself irrevocably to any theory whatever; and the one commandment of freedom was thus expressed, *Thou shalt live and not formulize.* That alone

was right which was done of one's own inner conviction
and mere motion; that was lifeless and evil which was
done out of obedience to any external authority.
'There is one thing in the world,' Clifford wrote about
this time, 'more wicked than the desire to command,
and that is the will to obey.' Now this doctrine of
individual and independent morality may look on the
face of it anarchical, and therefore it may be worth
while to observe that the Catholic doctrine of the duty
of following conscience is essentially at one with it.
The conscience may or may not be rightly informed.
It may be wrongly informed without one's own fault, as
in the case of invincible ignorance, or with it, as in the
case of culpable ignorance or perversity. But even in
this last case we are told that the sin of doing an absolutely wrong thing in obedience to the voice of conscience, however misguided, is infinitely less than the
sin of doing the absolutely right thing against one's conscience. The conscience must be rightly informed before
a completely right action is possible.[1] Again, Fichte
treats the sense of will and duty (from which he deduces
not only morality but the existence of other men and of
the world, in fact all knowledge and reality whatever)
as absolutely personal and individual. Clifford's early
doctrine of freedom was ardent and immature; but
whoever should call it immoral would find himself committed to applying the same language to some of the

---

[1] See the authorities collected in Dr. Newman's *Letter to the Duke of
Norfolk*, pp. 65, 66:—'Secundum sententiam, et certam, asserentem esse
peccatum discordare a conscientia erronea, invincibili aut vincibili, tenet D
Thomas, quem sequuntur omnes Scholastici.' 'In no manner is it lawful to
act against conscience, even though a law or a superior commands it.' Some
writers even say that this opinion is *de fide*.

greatest moralists of the world. The social theory of
morality stated and partly worked out in the ethical
portion of Clifford's essays is quite independent of this
earlier phase. At the same time it is not necessarily
inconsistent with it; for the determination of social
morality is apart from the assignment of motives for
individual morality, and leaves untouched the cultivation of individual perfection. Clifford, however, does in
his later writings freely and distinctly recognize the
validity of the social, or, as he sometimes calls it, the
tribal judgment, on the moral character of individual
acts regarded as an external quality; and there was a
time when he would probably have hesitated to allow
this.

In a note-book of Clifford's later Cambridge time
there are some speculations on the compensating intellectual pleasures that help to break the shock of
parting with old beliefs. I make an extract from one
of these pages. 'Whosoever has learnt either a language or the bicycle can testify to the wonderful
sudden step from troublesome acquirement to the mastery of new powers, whose mere exercise is delightful,
while it multiplies at once the intensity and the objects
of our pleasures. This, I say, is especially and exceptionally true of the pleasures of perception. Every
time that analysis strips from nature the gilding that
we prized, she is forging thereout a new picture more
glorious than before, to be suddenly revealed by the
advent of a new sense whereby we see it—a new
creation, at sight of which the sons of God shall have
cause to shout for joy.

'What now shall I say of this new-grown perception of Law, which finds the infinite in a speck of dust, and the acts of eternity in every second of time? Why, that it kills our sense of the beautiful, and takes all the romance out of nature. And moreover that it is nothing more than a combining and re-organizing of our old experiences, never can give us anything really new, must progress in the same monotonous way for ever. But wait a moment. What if this combining and organizing is to become first habitual, then organic and unconscious, so that the sense of law becomes a direct perception? Shall we not then be really seeing something new? Shall there not be a new revelation of a great and more perfect cosmos, a universe freshborn, a new heaven and a new earth? *Mors janua vitæ*; by death to this world we enter upon a new life in the next. A new Elysium opens to our eager feet, through whose wide fields we shall run with glee, stopping only to stare with delight and to cry, " See there, how beautiful ! " for the question, " Why ? " shall be very far off, and for a time shall lose its meaning.

' For a time ? It may well be that the new world also shall die. Doubtless there shall by and by be laws as far transcending those we know as they do the simplest observation. The new incarnation may need a second passion ; but evermore beyond it is the Easter glory.

Even at the time of these half-poetical meditations I think Clifford must have felt them to be too poetical for scientific use. Later in life, as we have seen above and may see in the Essays, he chose to make sure of a solid foundation in experience at the cost of sacrificing ornament and rhetoric, and his admiration of Mazzini

became compatible with practical empiricism in politics.
'On the whole I feel confirmed,' he wrote in a letter,
'that the English distrust of general principles in a
very complex affair like politics is a sound scientific
instinct, and that for some time we must go blundering
on, finding out by experience what things are to be let
alone and what not.'

The command, 'thou shalt not formulize,' was expressed in an amusing shape in a review of 'Problems
of Life and Mind,' published in 1874. 'Rules of philo-
sophizing are admirable things if two conditions are
satisfied : first, you must philosophize before you make
your rules ; secondly, you should publish them with a
fond and fervent hope that no philosophizer will attend
to them.'

As to Clifford's ideas on metaphysics proper I have
not much to say beyond what is disclosed in the Essays
themselves. His interest in philosophy grew up rapidly
after he took his degree, as is generally the case with
men who have any bent that way. I remember many
long talks with him on metaphysical questions, but not
much of the substance of them. One evening in the
Long Vacation of 1868, when we were up for the
Fellowship examination, we discussed the Absolute for
some couple of hours, and at last defined it to our own
exceeding content as that which is in necessary relation
to itself. Probably we laughed at our definition the
next morning, or soon after ; but I am still of opinion
that, as definitions of the Absolute go, this will do quite
as well as any other. Clifford's philosophical reading
was rather select than wide. He had a high admiration
for Berkeley, next only to Hume, and even more, perhaps,

for the 'Ethics' of Spinoza. The interpretation of
Spinoza's philosophy which I have put forward on one or
two occasions was common to Clifford and myself, and
on that subject (as, indeed, on everything we discussed
together) I owe very much to him. He was to have
lectured on Spinoza at the London Institution in 1877,
but his health would not allow it. There is little doubt
that this would have been one of his most brilliant and
original discourses. Students of Spinoza will easily
trace the connexion between his theory of mind and
matter and the doctrine set forth in Clifford's Essays on
' Body and Mind,' and 'The Nature of Things-in-themselves.' This was arrived at, to the best of my recollection, in 1871 or 1872 ; certainly before 1874, in
which year the last-mentioned paper was read at a
meeting of the Metaphysical Society. Briefly put, the
conception is that mind is the one ultimate reality ; not
mind as we know it in the complex forms of conscious
feeling and thought, but the simpler elements out of
which thought and feeling are built up. The hypothetical ultimate element of mind, or atom of *mind-stuff*,
precisely corresponds to the hypothetical atom of matter,
being the ultimate fact of which the material atom is
the phenomenon. Matter and the sensible universe are
the relations between particular organisms, that is,
mind organized into consciousness, and the rest of the
world. This leads to results which would in a loose
and popular sense be called materialist. But the theory
must, as a metaphysical theory, be reckoned on the
idealist side. To speak technically, it is an idealist
monism. Indeed it is a very subtle form of idealism,
and by no means easy of apprehension at first sight

Nevertheless there are distinct signs of a convergence towards it on the part of recent inquirers who have handled philosophical problems in a scientific spirit, and particularly those who have studied psychology on the physiological side. Perhaps we shall be told that this proves the doctrine to be materialism in disguise; but it is hardly worth while to dispute about names while more serious things remain for discussion. And the idea does require much more working out; involving, as it does, extensive restatement and rearrangement of metaphysical problems. It raises not only several questions, but preliminary (and really fundamental) problems as to what questions are reasonable. For instance, it may be asked why, on this hypothesis, mind should become conscious at a particular degree of complexity, or be conscious at all. I should myself say that I do not know and do not expect ever to know, and I believe Clifford would have said the same. But I can conceive some one taking up the theory and trying to make it carry further refinements and explanations. Again, a more subtle objection, but in my opinion a fallacious one, would be that it is not really a monism but a dualism, putting mind (as the undetermined *mind-stuff*) and consciousness in place of the old-fashioned matter and mind. This, however, is not the place to pursue the subject; and I do not think the outline of the hypothesis can be made clearer by any explanation of mine than Clifford has already made it.

After all I have wished to speak of the man rather than his opinions; but the speculative interests I shared with him, being in a manner part of himself, have claimed their due, and perhaps obtained rather more.

Let us now gather up a few matters of personal habit and character which have not yet been noticed. The predominance of light as a figure and a symbol in Clifford's writing will be remarked : he associates it with the right and all things good so constantly and naturally that it is one of the marks of his style. He had physically a great love of light, and chose to write, when he could, in a clear and spacious room, with the windows quite free of curtains. Though he was not for most ordinary purposes a business-like man, and was careless of his own attire, he was neat and exact in his literary work. He would not allow books to be misused or carelessly cut, and his own MS. was very fair, regular, and free from erasures. He was careful about punctuation, and insisted on having his own way in it, and he especially disliked superfluous commas. At the same time he was fond of handicraft, and his thoughts often ran upon mechanical invention. He speculated much on the practicability of constructing a flying machine, and began experiments at sundry times, which, however, never led to anything definite. Indeed it is pretty obvious that if a successful flying machine is ever made (and there is no impossibility in it), the inventor will be some one who combines theoretical knowledge of mechanics with familiar knowledge of machinery and the strength of materials and ready command of the various resources of engineering. At one time the notion of the flying machine turned Clifford's attention to kites, and this led to a ludicrous accident. It was in the Long Vacation of 1877, when Clifford and his wife were Mrs. Crawshay's guests in Wales. A kite of unusual dimensions, with tail in proportion, had been made ready for a flight which was to exceed everything achieved by kites before. It was to be flown with a great length of string, and it cost a morning's work to lay out the string in a field so that the kite might rise easily when started. Having accomplished this, the party went in to luncheon, and were presently called out by the announcement that a flock of sheep had been turned into the field. Clifford rushed out to prevent the disaster, but it was too late. Shepherd and sheep were caught as in a snare, and when they were extricated the string was left hopelessly entangled. Another piece of engineering undertaken at the same time and place was the construction of a duck-pond for the benefit of a family of ducklings who frequented a narrow ditch by the roadside. The little stream that trickled in the ditch was dammed according to the rules of art, and in course of time a complete pond was formed, and the ducks were happy for a season: till one day some over-zealous minister of local authority, conceiving the pond, as it was supposed, to be an encroachment on the highway, restored the ancient state of things with a few strokes of the spade. Clifford regretted the duck-pond even more than the kite. Other amusing and characteristic anecdotes might be added; but I forbear.

No enumeration of tastes and occupations can adequately represent the variety and flexibility of Clifford's intellect, and still less the tender, imaginative, poetical side of his mind. Now and then he wrote verses in which this partly found expression. They were mostly of a private or occasional nature, or else too fragmentary for publication. One very graceful song is to be found in the volume of fairy tales already spoken of, and is reprinted below. But the real expression of Clifford's varied and fascinating qualities was in his whole daily life and conversation, perceived and felt at every moment in his words and looks, and for that very reason impossible to describe. Nor can portraits go very far to supply that part of it which fell to the sight; for the attractive animation and brightness of his countenance depended on very slight, subtle, and rapidly succeeding changes. His complexion was fair ; his figure slight, but well-knit and agile; the hands small, and, for a man, singularly slender and finely formed. The features were of a massive and irregular type which may be called Socratic ; in a bust they might have looked stern, in the living face they had an aspect not only of intellectual beauty but of goodwill and gentle playfulness. But I began with declaring my task impossible, and at the end I feel still more keenly that all words fall short of what I would convey. The part has fallen to me of doing to a loved and honoured friend such honour as I could : the will at least will be accepted.

> Purpureos spargam flores . . et fungar inani munere.

PART II.

SELECTIONS FROM LETTERS, ETC.

THE following is a selection from letters written by Clifford at various times, partly to my mother and partly to myself. I begin with some philosophical passages.

[*To F. Pollock.*]

'Trinity College, Cambridge: April 2, 1870.

' Several new ideas have come to me lately : first, I have procured Lobatschewsky, " Études Géométriques sur la Théorie des Parallèles " . . . a small tract of which Gauss, therein quoted, says, " L'auteur a traité la matière en main de maître et avec le véritable esprit géométrique. Je crois devoir appeler votre attention sur ce livre, dont la lecture ne peut manquer de vous causer le plus vif plaisir." It is quite simple, merely Euclid without the vicious assumption, but the way the things come out of one another is quite lovely. . . .

' I am a dogmatic nihilist, and shall say the brain is conscious if I like.' (This in reply to some verbal criticism of mine.) ' Only I do not say it in the same sense as that in which I say that *I* am conscious. It seems to me that not even Vogt, however you fix it, can talk about matter for scientific purposes except as a phenomenon ; that in saying the brain is conscious—or, better, that *you* are conscious, I only affirm a correlation of two phenomena, and am as ideal as I can be; that, consequently, a true idealism does not want to be stated, and, conversely, an idealism that requires to be stated must have something wrong about it.   In the same way to say that there is God apart from the universe is to say that the universe is not God, or that there is no real God at all; it may be all right, but it is atheism.   And an idealism which can be denied by any significant aggregation of words is no true idealism.'

The following is on the recent edition of Hume by Messrs. Green and Grose :—

[*To F. Pollock.*]

'Exeter: September 11, 1874.

'. . . I hope you have seen Sidgwick's remarks (I think in the " Academy ")[1] ; he points out that to prove Hume insufficient is not to do much in the present day. It should I think be brought out clearly that if we pay attention only to the scientific or empirical school, the theory of consciousness and its relation to the nervous system has progressed in exactly the same way as any other scientific theory ; that no position once gained has ever been lost, and that each investigator has been able to say " I don't know " of the questions which lay beyond him without at all imperilling his own conclusions. Green, for instance, points out that Hume has no complete theory of the *object*, which is of course a very complex thing from the subjective point of view, because of the mixture of association and symbolic substitution

[1] May 30, 1874, vol. v., p. 608.

in it ; and in fact I suppose this piece of work has not
yet been satisfactorily done.   But it seems merely per-
verse to say that the scientific method is a wrong one
because there is yet something for it to do ; and to find
fault with Hume for the omission is like blaming Newton
for not including Maxwell's Electricity in the Principia.'

The following suggestions on education were sent
from Algiers in June, 1876 :—

[To F. Pollock.]

'. . . I have a scheme which has been communicated
in part to Macmillan and which grows like a snowball.
It is founded on " Pleasant Pages," the book I was taught
out of; which is a series of ten minutes' lessons on the
Pestalozzian plan of making the kids find out things
for themselves : history of naughty boys on Monday,
animals on Tuesday, bricks on Wednesday, Black Prince
on Thursday, and so on.   In the book it was very well
done, by a man who had a genius for it.   If you go to see
Macmillan in Bedford Street he will show you the book,
which he got on my recommendation—he is also himself newly interested in the question.   His partner Jack
read part of it and was struck.   Well, I first want that
brought up to to-day, both in choice of subject and in
accuracy ; adding, e.g. a series of object lessons on man
(papa, mamma, house, street, clothes, shop, policeman,
" wild and field ").   Then I want it taught on the
Russian system, in different languages on successive
days ; no direct teaching of language until there are
facts enough to make Grimm's law intelligible, for which
English, German and the Latin element in French would
be enough ; no grammar at all till very late, and then

as analysis of sentences and introductory to logic. This is the difficult part; it would require a French and German teacher, both trained and competent, besides the English one. So far as the book is concerned it would of course be easy to print it in the three languages. Lastly, I have bought twelve volumes of the Bibliothèque Nationale for three francs—Rabelais, five volumes, and Montesquieu, Pascal, Diderot and Vauvenargues. They are twenty-five centimes each, admirable for the pocket —and of course you know them. There are two or three hundred volumes. Whereupon we must of course get the same thing done for English literature, and the setting forth of all literature in English (e.g. I have " Les Maximes d'Epictète "), but more particularly we must get published excellent little manuals at twopence or three-pence for the use of Board and other primary schools. I do not even know that penny school books would not be a successful move—the size of a " Daily News," say, printed by the million in a Walter press, folded and sewed by machinery to about the size of the Bibliothèque.

' A " Daily News " would just make one of these volumes. Fancy the " Pensées " of Pascal, with the notes of Voltaire, Fontenelle, and Cordorcet, a good life at the beginning, etc., all well printed on a sheet of the " Daily News! " But of such a size could be made a very good elementary schoolbook of arithmetic, geometry, animals, plants, physics, etc.—rather larger than Macmillan's primers, but of the same sort.'

The remaining letters and extracts are chiefly de-scriptive, and will be given without further remark, except such brief note of dates and circumstances as may seem necessary.

[To Lady Pollock.]

'Cambridge, September 26, 1871.

'. . . Now I shall confess that on two occasions I have wasted time lately. One was due to A. who, seeing my rooms [in London] empty, and with waxed floors (to save carpets), insisted on bringing three or four of her friends to dance there . . . The weather was still comparatively warm and we could use the balcony. I chiefly remember a waltz Eberlein played us, and which really made me believe in the existence of a tarantelle. It was just like being in a high surf of the sea. You thought, as you reached the wall panting and helpless, that there might be at last a moment's peace; but then there came a crash in the music like the breaking of a wave, and away you were swept into a tumult of fury, the strength of which after all seemed to be your own. I was dancing with A. herself, who really goes wonderfully well and was then inspired. The worst is that Eberlein can't recollect what it was, or whether he improvized it or anything about it; so that it will just remain as a memory of what is possible . . . Crotch and I went down to T., where we showed off our somersaults to C.'s utter consternation, and generally amused everybody. Also an odd thing occurred when C. was trying to drive me behind a pair of horses; I suddenly found myself going alone at a furious pace into the High Street of Cheltenham; dived over the splashboard for the reins, and was just beginning to think the situation interesting, when the near horse kicked over the crowbar into the spokes of the wheel exactly as the other one pulled up against a lamp-post. My C., who had been pitched elegantly on to his head, came up, brushing his hat with the smiling remark, "By Jove! I thought that was the end of you;" we then impounded a man with a saw to cut the horse free, and went on after twenty minutes' delay consumed in getting another trap . . .

'My ideal theory is quite different from yours. In the case of persons I worship the actual thing always; this is the only way to be trusted. The one advantage of having indestructible family relations is that, whatever you do and whatever anybody thinks of you, there are always one or two people who will love you exactly as much as (if not more than) if you were blameless and universally respected. I used to recognize an exception, viz., that in certain cases what had been a person might cease to be one, and become a thing, towards which one could have no moral relations, and which might be set aside by safe means, or used as the occasion served. But the more people I know and the better I know each, the further off this possibility seems to be. I want to take up my cross and follow the true Christ, humanity; to accept the facts as they are, however bitter or severe, to be a student and a lover, but never a lawgiver. But then besides this I do look for an ideal which is at some time to be created or awakened out of potentialities—like the lady that Phantastes set free from the block of marble. Meanwhile I chip various blocks, and generally set free something; not hitherto I think quite the right one; when I do she will probably go straight off to somebody else. All this, by the way, is only theory; my practice is just like other people's.'

[*To Lady Pollock.*]

' Bagnères de Bigorre, Hautes Pyrénées: Summer of 1870.

' I really don't know what day, except that there has been a *dimanche* lately; we came home late from the mountains and found a large crowd saying " Voilà le feu d'artifice ! " so we concluded it was Sunday. But then that was before we had been anywhere or done anything, and I was really ashamed to write. This, (here the letter proper begins), is the result of our casual way of taking things; we are still in France— not that we have not been to Spain—but we have never reached Santander, which was to have been the first halt. At Havre there was a quarter of an hour to spare, so I return from a stroll in the town laden with strawberries, cherries, and apricots, just in time to reach the already-started boat by one of those apparently unpremeditated springs which look so well in the Gymnasium. At Honfleur a surprising meal, *bouillon, côtelettes, vin*— till we were roaring drunk—for sevenpence-halfpenny each. Then various towns in Normandy, which I have hopelessly mixed up. " Lisieux était—on ne peut plus s'imaginer—délicieux." This is Crotch's [1] abominable pun. There was a fair at Le Mans, and we nearly broke the merry-go-round. At Tours we caused two mild priests *faire signe de croix*, by suddenly flapping " Le Rappel " and " La Liberté " from our bag on the ramparts. But Angoulême ! everybody must go there at once and stay several years. It is too lovely. You walk under trees all round on the top of the walls, and

---

[1] A brother of G. Crotch, who has been already mentioned in the biographical part of this Introduction.

see miles of Garonne and vines.   It was *fête-Dieu* about
the time we got to Bordeaux and Bayonne, and all the
little dears were in white for their first communion.
Then came troubles; there were no boats.   We got
by rail to San Sebastian, which indeed is sweetly
pretty; so that I was moved even to try to sketch the
Plaza Reale—such is the audacity of some.   But there
was still no boat to Santander; the diligence was out
of the question, and the only way by rail was to go to
Burgos and back again; a proceeding apparently
ridiculous, as the same ground would have to be re-
traced, by me at least, in going to Madrid.   So we decided on letting the other people take care of themselves, and doing our Pyrenees on the French side
instead of the Spanish.   Perhaps you know this place;
it is said to be the prettiest and cleanest town in France,
and I know nothing to the contrary.   By the way we
spent a day at Biarritz and a day at Pau, on our way
here; the former is rather like Ilfracombe, but I don't
think much of it.   Pau is called here *le petit Paris*, a
judgment doubtless comparative; it is certainly not
bad.   Here we have hired a garret near the sky,
and live charmingly on five francs a day; this is
accomplished chiefly by getting a cheap bed and not
eating anything.   All day long we catch butterflies
and sketch.   Sometimes we go to a *table-d'hôte*; where,
besides the ordinary fare, Crotch finds *sortie du flanc
d'Adam, côtelette funeste*—a young lady who won't speak
to me when he is by.   But our great adventure is the
Pic du Midi; close to the top of which is an *hôtellerie*,
containing (in the guide-book), ham and eggs, with
people to cook them, but (in fact and at this time of

year) not a living soul, and only three inches of candle.
However, we broke in at a window and made a fire;
also we shouted to a shepherd whom we persuaded to
bring us some milk, then quarrelled with him because
he wanted more than two francs, so that he went away
swearing we should not sleep quietly. But that we did
after a walk of over twenty miles, with a rise of 10,000
feet; notwithstanding that it was cold, and that a most
elegant and meritorious bear came and sniffed all round
in the night. The great point, of course, was the sunset and sunrise at the top. We were rather disap-
pointed not to see the Atlantic at the former, which is
occasionally done, but we saw the whole range of the
Pyrenees, which is an institution not to be despised.
Goat's milk half-way down, trout at Grippe, Bagnères
in the evening; general astonishment and increased
respect of the natives. They won't hear of my going
to Spain—*que c'est un pays affreux, et qu'on m'égorgera*
—which is amusingly improvident from Frenchmen.
What pleased me most was to see in a Spanish railway-
carriage, in pencil on the wall:

> " Qui no quiere rey es mi amigo
> Viva la republica federal."

I have just read Madame Thérèse (Erckmann-Chatrian),
it is a good book. Roi des Montagnes is capital; I shall
never quite know why brigand-shooting is not a
favourite pastime. Apollo, you will be glad to hear, is
the finest butterfly in Europe, white, with red spots;
caught five of them. Also a dear little red viper in my
butterfly net; carried him some miles, then got tired
and squashed his head with a stone; they kill horses

sometimes.   I have nice new shoes of straw, 2 fr. 50 cents., on the Pic du Midi I looked despairingly at these, and said, " This flesh also is grass."   But we had some chocolate and two biscuits with us, so we spared even Crotch's boots.   The day after to-morrow I get to Burgos; then to Madrid for a week; then to some other university town.   Crotch returns by Hamburg and Copenhagen to Norway.   Nobody ever dies here; they get smaller and smaller, and are ultimately kept in pill-boxes.   It is thought pious to keep a microscope and look at your gr—gr—gr—gr—grandfather.   However we have tried the mineral waters, and don't believe in them.   Beso los piés de Vd., y estoy á servirla.'

[To Lady Pollock.]

' Florence, December, 1870.

(Clifford was one of the English Eclipse expedition : the *Psyche*, with the expedition on board, struck on a rock near Catania.   All hands and the instruments were saved, the ship was lost.)

' No ink, no paper, no nothing—Florence, Thursday 5th.   The above[1] you guess.   After that somehow to Catania, some in boats and some in holy carts of the country, all over saints in bright shawls—well, if ever a shipwreck was nicely and comfortably managed, without any fuss—but I can't speak calmly about it because I am so angry at the idiots who failed to save the dear ship—alas ! my heart's in the waters close by Polyphemus's eye, which we put out.   At Catania, orange groves and telescopes; thence to camp at

---

[1] A grotesque fancy sketch of the shipwreck.

Augusta; Jonadab, son of Rechab, great fun, natives kept off camp by a white cord; 200 always to see us wash in the morning—a performance which never lost its charm—only five seconds totality free from cloud, found polarization on moon's disk, agree with Pickering, other people successful. Then by Catania to Messina, no steamers, kept five days, Mediterranean stormy, we also at last to Naples, very bad night, everybody ill but me, and I have been out of sorts ever since. Called on Mrs. Somerville, and came on to Rome after seeing Pompeii. At Rome $2\frac{1}{2}$ days, pictures, statues, Coliseum by moonlight. Both of us sneezed awfully next morning. The shops are in the streets where the Tiber left them—nice for purchasing but not so convenient for walking about. This morning arrive in Florence—Pitti palace—spent all my money, and shall get stranded between Cologne and Ostend unless I can live on one egg every other day, and thereout suck no small advantage,—be better off in Paris. Addio.'

[To Lady Pollock.]

'Sunday, July 2, 1876.

'This comes from Oran in the west of Algeria, a sad place, with too many Spaniards in it. We came here yesterday after a long and tiresome journey from Blidah, near Algiers. The train is somewhat amusing because the carriages are open at the ends and you can sit in the air as if it was a tram-car. You have then to be careful not to let the very large grasshoppers eat you up. Playfair, the English Consul at Algiers, told us to go to Bougie to see the gorge of the Chabet; so we got a Murray's Guide and started off obediently. It was the steamer that had brought us from Marseilles, and the captain, who is very fond of us, gave us the ladies' cabin all to ourselves. There was on board a little Frenchman who had observed us in a restaurant at Algiers. He made great love to us, and said he wanted to marry an Englishwoman, but we think he lied a good deal about his town and country house, and his carriage and his good family. However, he woke us up in time for the diligence at Bougie, and there is no harm in him, though indeed very little else. All this expedition was undertaken for the sake of the road from Bougie to Sétif, and it was well worth it. There is a narrow rent made by the stream which winds in and out for miles among the hills; these are splendidly wooded, and rise to an enormous height on either side, while the torrent roars away down below. The road is cut in one side of the gorge. The cochon who drove the diligence tried every ruse to get us inside, that he might have a friend of his on the front seat; but we stuck to our places till the scenery was finished, and then a great rain came and drenched both of them well. Sétif is a complete French town, stuck in the middle of an African plain with its cafés and boulevards, just as if it had never lived anywhere else. We saw more Arabs there than anywhere else, and the native market pleased us much. On the way back we travelled with an Arab who had a gazelle in a basket which he was taking to somebody at Bougie; he said you might buy them occasionally in the market at Sétif for twenty-five francs: we pitied the sweet little thing, which baaed like a sheep and struggled hard to get out, but he was pacified with some bread and some flowers which I had

picked, and went to sleep with his head on my arm.
On waking up he saw Lucy's straw hat near him and
tried to eat it. We saw the most exquisite masses of
maiden-hair fern, as large as the side of a room (the
masses I mean, not the fern), where the streams came
down near the side of the road. Our little Frenchman
was still at Bougie and came back with us in the boat.
The next day but one we had an amusing experience in
the Jardin d'Acclimatation. We were taking coffee in
an Arab café, and there was a boy there with an instrument of two strings, whose sounding board was made
of bladder stretched over the shell of a tortoise—quite
the Apollo. We asked him to play something to us,
and then a flute painted red and blue was given to an
old man who had been smoking quite still. I couldn't
make out the music because the little Frenchman kept
on chattering ; but the old man gradually became excited ; he had been sitting European fashion with his
feet on the ground, but one of his great toes got restive
and then all the others, until his shoe was too much for
that foot ; so he dropped the shoe and laid the foot on
his knee, where it could wriggle comfortably. Then the
other foot became excited and went through the same
process. When his agony grew still more intense, he
put one foot down and bent the shoe about with it to
get more resistance. All this time the upper part of
his body, except the fingers playing on the pipe, was
perfectly still, and his face had a rapt expression.
Meanwhile a pipe of *kif* had been got ready and was
handed round, and a whiff of that seemed to calm him.
I tried it also, and it brought the tears into my eyes, I
was so nearly suffocated. I went to a lecture of the

Arabic course which is given at Algiers in the Museum. It consisted in the translation of an article from a Constantinople paper, passages from which were written up on a black board, read out, and translated. The point of interest was the quotation from a passage in the Koran in support of the constitution, to the effect that 'the government shall not be absolute but consultative.' The lecturer said that absolutism was a Turkish institution, not Arabic, and that the Caliphate had been a sort of republic, with a president elected for life. Also that when a certain Caliph boasted that he had never swerved from the path of justice, a soldier looked up and said 'Inshallah! (or words to that effect, meaning, By Jove!) our swords would have speedily brought you back.' This appears interesting if true. Already a Parisian scent is sold in the Moorish bazaars as a perfume of the Sultana Valide.

'We felt very much injured at only seeing two monkeys in the woods at La Chiffa the day before yesterday, but there were some green parrots on the bushes near the railway.

'To-morrow we go by a Spanish boat to Almeira, and thence by diligence or another boat to Malaga. The Spanish boat will be nasty, but it is only twelve hours or so. I am very much better, and shall be glad of a rest at Granada after this gadding about.

'P.S. I wrote to Fred about the education of our infants. I am very glad we have both begun with girls, because it will be so good for the other children to have an elder sister. How very fond those kids will be of each other and of Fred and me! because girls always like their fathers best, you know. I have thought of a way to make them read and write shorthand by means of little sticks (not to whop them with but to put together on a table and make the shorthand signs). Ask G. whether she thinks they had better learn to sing on the sol-fa system; it is very amusing and seems to me more adapted for children than the other. Of course I can teach them to stand on their heads.

'We have seen the Spanish boat, which is called *La Encarnacion*, and that rightly ; for it is the incarnation of everything bad.'

[The *Encarnacion* aforesaid more than justified the worst expectations : the engines broke down at sea, nobody on board was competent to repair them, and the ship lay helpless till a vessel was hailed which had a French engineer on board.]

[*To F. Pollock.*]

'Malaga, Saturday, July 15, 1876.

'. . . As for this country, I think it requires to be colonized by the white man. The savages would gradually die out in his presence. The mark of a degraded race is clear upon their faces ; only the children have a look of honesty and intelligence, a fact which is also observed in the case of the negro, and is a case of Von Bär's law, that the development of the individual is an epitome of that of the race. It is instructive also to contrast the politeness fossilized in their language with the brutal coarseness of their present manners, of which I may some time tell you what I will not soil paper with. I think it possible that one Spaniard may have told me the truth : he had lost so many teeth that he left out all his consonants, and I could not understand a word he said. When we went on board the *Rosario* at 11 p.m. the boatmen stood in the way to keep us from the ladder, and threatened us for the sake of another peseta over the regular charge. The steward tried to cheat me over the passage-money, but I appealed to the authorities who came on board at Malaga and got the money back (there are many strangers here). Then he made another grab in the matter of our breakfasts, in the face of a tariff hung up in the cabin. It is tiring to have to think that every man you meet is ready to be your enemy out of pure cussedness. I don't understand why one is expected to be polite and reticent about the distinction between the Hebrew piety and Roman universalism attributed to Jesus and Paul, and the ecclesiastical system which is only powerful over men's lives in Spain, the middle and south of Italy, and Greece —countries where the population consists chiefly of habitual thieves and liars, who are willing opportunely to become assassins for a small sum. I suppose it frightens people to be told that historical Christianity as a social system invariably makes men wicked when it has full swing. Then I think the sooner they are well frightened the better.'

[*To F. Pollock.*]

'Washington Irving Hotel, Granada, August 3, 1876.

'You are quite right, and one ought not to despair of the republic. These folks are kind and rather pleasant when one is *en rapport* with them, and they have a deal of small talk. We found a jolly old couple one morning when we were coming back from a hot walk in the Vega of Almeira (*vega* = cultivated plain surrounding a town which feeds it) ; we asked for some milk, which they had not, but they gave us a rifresco of syrup and cold water, not at all bad, and the old woman showed Lucy all over her house while the man smoked a cigarette with me. Lucy's passport is the baby's portrait, with which she gains the hearts of all the women and most of the men. What made it more surprising was that they took us for Jews. Wilkinson, our Consul at Malaga, who has been here with his wife and daughter (awfully nice people and cheered us up no end), says that the country people are better than those in the towns.

'. . . . But although we have been nearly a fortnight at Granada, only one murder has been even attempted, so far as I know, within 100 yards of the hotel. A. had been making love to B.'s wife, and so she was instructed to walk with him one evening under these lovely trees. She took occasion to borrow his sword-stick, and stuck him in the back with it while her husband fired at his head with a revolver. One ball grazed his temple, and another went in at his cheek and out of his mouth, carrying away some teeth and lip. He came round to the Spanish hotel opposite and was tied up on the doorstep; they dared not let him come in because the police are so troublesome about these affairs. The defence was that A. was a Republican, and had been a Protestant; so you see B.'s love of order was such that he did not think jealousy a sufficient justification. Wilkinson had just received a report of the last quarter of 1875 ; in those three months there had been only a few more than 400 murder cases in the whole province of Granada. The hot weather seems to try them; a paragraph in the Malaga paper, headed ' Estadístico Criminal de Domingo, 30,' gives fifteen cases of shooting and stabbing last Sunday in Malaga, but only five appear to have been fatal. This is not assassination, but is merely an accompaniment of their somewhat boisterous conviviality; they get drunk together and then draw their knives and go in for a hacking match. It is not even quarrelling in all cases; in Granada the other day three men shut themselves up and fought till they were all dead. They might, to be sure, have disliked each other mutually all round, but I am inclined to think it was a party of pleasure rather than of business They do not attack strangers in this wav (i.e. with knives and revolvers), unless, of course, there is a reason for it; but when anything offends their delicate sense of propriety one cannot expect them not to show it a little. Thus they threw stones in Seville and Cordova at a lady who is now staying here, because she went into the street by herself, and they do not approve of that. I am afraid my Norfolk jacket hurts their feelings in some way, but they have been very forbearing, and have only stoned me once, and then did not hit me. Another time a shopkeeper set his dog at me, but although this was rather alarming, with temperature 92° in the shade, it must have been meant as a joke, for Spanish dogs only bite cripples of their own species—except, indeed, the great mastiffs that are kept to bait bulls that won't fight. Of course one is not so insular as to think there is only one way of giving a welcome to the stranger; and the ' 'eave 'arf a brick at 'im ' method is improved by variety. What generally happens is this : the grown people stop suddenly at the sight of you, and wheel round, staring with open mouths until you are out of sight; while the children, less weighted with the cares of this world, form a merry party and follow at your heels. When you go into a shop to buy anything, they crowd round the door so that it is rather difficult to get out. The beggars come inside and pull you by the arm while you are talking to the shopman. I have invented a mode of dealing with the crowd of children; it is to sit on a chair in the shop door and tickle their noses with the end of my cane. I fear that universal sense of personal dignity which is so characteristic of this country is in some way injured by my familiarity; the more so as it cannot be resented, for the other end of my cane is loaded, and I do not try it on in a macadamized street. Anyhow they go a little way off. In Malaga the people seemed more accustomed to the sight of strangers, and contented themselves with shouting abusive epithets. . . Everybody says there will be a revolution before long. . . If Castelar returns to power, I hope among other little reforms that he will prevent the post-office officials from stealing letters for the sake of the stamps on them; it is a great interruption to business and must be a laborious way of earning money. One of them was caught in Malaga because a packet of letters which he had thrown into the sea was accidentally fished up; but he was shielded from punishment by the authorities.

'We are very happy here, with a Swiss cook and an Italian landlord. There are some English, Germans, and Italians staying over the way, and in a few minutes we can be among the memorials of a better time. I am too tired now to talk about the Alhambra, but it seems to me

to want that touch of barbarism which hangs about all
Gothic buildings. One thinks in a Cathedral that since
somebody has chosen to make it it is no doubt a very
fine thing in its way; but that, being a sane man, one
would not make anything like it for any reasonable
purpose. But the Alhambra gives one the feeling that
one would wish to build something very like it, *mutatis
mutandis*, and the more like it the more reasonable the
purpose was. Moreover, I think it must be beautiful,
if anything ever was; but then I have no taste.'

From ' *The Little People.*'

1.—SONG.

THIS is the song that Daisy sang; and it is about a
water-lily bud that saw a reflection of herself in the
surface of the water while she was under it.

> You grow through the water apace, lily;
>   You'll soon be as tall as the pond,
> There is fresh hope high in your face, lily,
>   Your white face so firm and so fond.
>   Ah, lily, white lily,
>   What can you see
>   Growing to meet lily
>   Graciously?

> There's a face looks down from the sky, lily;
>   It grows to me dim from above.
> If I ever can reach me so high, lily,
>   I shall kiss—ah! the face of my love.
>   Ah, lily, white lily,
>   That can I see,
>   Giving me light, lily,
>   Lovingly.

The lily-bud met with her mate, ah me !
And her flower came through to the air,
And her bright face floated in state, ah me !
But the shadow-love never was there !
Ah, lily, great lily,
Queenly and free,
Float out your fate, lily,
Friendlessly.

2.—THE GIANT'S SHOES.

ONCE upon a time there was a large giant who lived in a small castle ; at least, he didn't all of him live there, but he managed things in this wise. From his earliest youth up, his legs had been of a surreptitiously small size, unsuited to the rest of his body ; so he sat upon the south-west wall of the castle with his legs inside, and his right foot came out of the east gate, and his left foot out of the north gate, while his gloomy but spacious coat-tails covered up the south and the west gates ; and in this way the castle was defended against all comers, and was deemed impregnable by the military authorities. This, however, as we shall soon see, was not the case, for the giant's boots were inside as well as his legs ; but as he had neglected to put them on in the giddy days of his youth, he was never afterwards able to do so, because there was not enough room. And in this bootless but compact manner he passed his time.

The giant slept for three weeks at a time, and two days after he woke his breakfast was brought to him, consisting of bright brown horses sprinkled on his bread and butter. Besides his boots, the giant had a pair of shoes, and in one of them his wife lived when she was at home; on other occasions she lived in the other shoe. She was a sensible, practical kind of woman, with two wooden legs and a clothes-horse, but in other respects not rich. The wooden legs were kept pointed at the ends, in order that if the giant were dissatisfied with his breakfast he might pick up any stray people that were within reach, using his wife as a fork. This annoyed the inhabitants of the district, so that they built their church in a south-westerly direction from the castle, behind the giant's back, that he might not be able to pick them up as they went in. But those who stayed outside to play pitch-and-toss were exposed to great danger and sufferings.

Now, in the village there were two brothers of altogether different tastes and dispositions, and talents and peculiarities and accomplishments, and in this way they were discovered not to be the same person. The elder of them was most marvellously good at singing, and could sing the Old Hundredth an old hundred times without stopping. Whenever he did this he stood on one leg and tied the other round his neck to avoid catching cold and spoiling his voice; but the neighbours fled. And he was also a rare hand at making guava dumplings out of three cats and a shoe-horn, which is an accomplishment seldom met with. But his brother was a more meagre magnanimous person, and his chief accomplishment was to eat a waggon-load of hay overnight, and wake up thatched in the morning.

The whole interest of this story depends upon the fact that the giant's wife's clothes-horse broke in con-

sequence of a sudden thaw, being made of organ-pipes.
So she took off her wooden legs and stuck them in the
ground, tying a string from the top of one to the top
of the other, and hung out her clothes to dry on that.
Now this was astutely remarked by the two brothers,
who therefore went up in front of the giant after he
had had his breakfast. The giant called out, 'Fork!
fork!' but his wife, trembling, hid herself in the more
recondite toe of the second shoe. Then the singing
brother began to sing; but he had not taken into account the pious disposition of the giant, who instantly
joined in the psalm, and this caused the singing brother
to burst his head off, but, as it was tied by the leg, he
did not lose it altogether.

But the other brother, being well thatched on account of the quantity of hay he had eaten overnight,
lay down between the great toe of the giant and the
next, and wriggled. So the giant, being unable to bear
tickling in the feet, kicked out in an orthopodal
manner; whereupon the castle broke and he fell backwards, and was impaled upon the sharp steeple of the
church. So they put a label on him on which was
written 'Nudipes Giganteus.'

That's all.

PART III.

BIBLIOGRAPHICAL.

It seems desirable to give under this head a list, as complete as the Editors have been able to make it, and arranged in order of time, of Clifford's non-mathematical lectures and writings ; as well as his own scheme—which unfortunately must remain unexecuted—of recasting and consolidating his work. Those pieces which do not appear in this collection are named in italics. Several of the best lectures, it will be noticed, were first given for the Sunday Lecture Society, of which Clifford was a warm supporter. The object of the Society, namely the spreading of exact knowledge, and the treatment of Science, History, Literature and Art, with special regard to 'their bearing upon the improvement and social well-being of mankind,' was one thoroughly congenial to him ; and his aid may claim an appreciable share in the success that has hitherto attended the Society's operations. The words, 'No report,' mean that neither any MS. nor any sufficient report of the lecture, or paper published or unpublished, has come to the hands of the Editors.

'Conditions of Mental Development.' Royal Institution, March 6, 1868 (printed in Proceedings).

'Theories of Physical Forces.' Royal Institution, February 18, 1870 (printed in Proceedings).

'Aims and Instruments of Scientific Thought.' Address to British Association at Brighton, 1872: reported and reprinted in Macmillan's Magazine, October 1872.

'*The History of the Sun: being an explanation of the nebular hypothesis and of recent controversies in regard to the time which can be allowed for the evolution of life.*' Sunday Lecture Society, April 16, 1871 ; repeated at Exeter some time afterwards. There has come to our hands a MS. report taken at Exeter, which is unhappily so confused and imperfect that it has been found impossible to reproduce the lecture from it with anything like reasonable certainty. This lecture, therefore, remains unpublished.

'Atoms.' Sunday Lecture Society, January 7, 1872 ; repeated at Manchester, November 20, 1872, and printed in the series of Manchester Science Lectures.

'*Ether ; the Evidence for its Existence and the Phenomena it explains.*' Sunday Lecture Society, April 14, 1872. No report. Part of this appears to be substantially repeated in 'The Unseen Universe.'

'*Ultramontanism.*' Paper read before the London Dialectical Society, April 28, 1875. No report.

'*The Dawn of the Sciences in Europe.*' Sunday Lecture Society, November 17, 1872. No report.

'*The Relations between Science and some Modern Poetry.*' Sunday Lecture Society, May 4, 1873. Recast and enlarged as 'Cosmic Emotion.'

'Philosophy of the Pure Sciences.' Afternoon lectures at Royal Institution, March 1, 8, 15, 1873. Substance printed in Contemporary Review and Nineteenth Century, October 1874, February 1875, March 1879, thence reprinted here, together with a MS. addition.

Review of Vol. I. of G. H. Lewes' '*Problems of Life and Mind.*' Academy, February 7, 1874.

'The First and the Last Catastrophe.' Sunday Lecture Society, April 12, 1874. Printed in Fortnightly Review, April 1875, and also by the Sunday Lecture Society.

'*On the Education of the People.*' Royal Institution, May 22, 1874. Only short abstract in Proceedings. Repeated at Midland Institute, Birmingham, enlarged and divided into two lectures, February 1875. MS. report of latter part only, hardly intelligible without the diagrams frequently referred to.

'Body and Mind.' Sunday Lecture Society, November 1, 1874; Fortnightly Review, December 1874; also printed by the Society.

' On the Nature of Things-in-Themselves.' Paper read before the Metaphysical Society in 1874 ; published in Mind, January 1878.

'Seeing and Thinking.' Three lectures at Shoreditch for a University Extension Course, December 1874. These are to be published as a separate little book, according to the original intention, at or about the same time as the present volumes.

' The General Features of the History of Science.' Four afternoon lectures at Royal Institution, February 27, March 6, 13, 20, 1875. Only brief syllabus, and partial report of one lecture from shorthand notes.

' On Babbages' Calculating Machines.' Royal Institution, May 24, 1872. No report.

' The Unseen Universe.' Fortnightly Review, June 1875.

' On the Scientific Basis of Morals.' Paper read before the Metaphysical Society, 1875; Contemporary Review, September 1875.

'Right and Wrong.' Sunday Lecture Society, November 7, 1875 ; Fortnightly Review, December 1875, also printed by the Society.

' Sight, and what it tells us.' Lecture at London Institution, February 24, 1876. Partly to same effect as 'Seeing and Thinking.' Short report in Times, reprinted in the Lecture Supplement to the Journal of the London Institution. We are indebted for this information to the kindness of the Librarian, Mr. E. B. Nicholson.

'Instruments used in Measurement; Instruments illustrating Kinematics, Statics, and Dynamics. In the South Kensington Handbook to the Special Loan Collection of Scientific Apparatus,' 1876.

' Ethics of Belief.' Paper read before the Metaphysical Society, 1876 ; published with considerable additions in Contemporary Review, January 1877.

'Ethics of Religion.' Sunday Lecture Society, March 4, 1877; then entitled '*The Bearing of Morals on Religion:*' printed by the Society; and in the Fortnightly Review, July 1877, under the title now given, which Clifford intended to be the permanent one.

'The Influence upon Morality of a Decline in Religious Belief.' Nineteenth Century, April 1877, in a 'Modern Symposium.'

'Cosmic Emotion.' Nineteenth Century, October 1877.

'Virchow and the Teaching of Science.' Nineteenth Century, April 1878.

'*Childhood and Ignorance: a reason for not replying.*' Nineteenth Century, May 1878. A trenchant exposure of elementary blunders in physics committed by a pretentious critic: considered too short and occasional for republication.

It is possible that there may exist reports unknown to the Editors, in local journals or elsewhere, of some of the unpublished lectures. If there are any such, the Editors or the Publishers will be thankful for information of them.

In reprinting the Lectures and Essays which now appear no alterations have been made beyond such little matters of verbal and literary correction as the author would have naturally attended to if he had himself undertaken a revision with a view to collected publication; but certain passages have been omitted which we believe that Clifford himself would have willingly cancelled, if he had known the impression they would make on many sincere and liberal-minded persons whose feelings he had no thought of offending.

A few footnotes have been added for various reasons, which are distinguished from the author's own by being enclosed within square brackets. Some repetitions will be found on a comparison of different pieces, as might naturally be expected in discourses on kindred subjects composed without reference to one another. We have not made any attempt to remove these: partly because it is not easy to decide if either, and if so which, of any two parallel passages can be safely treated as superfluous with regard to its own context; partly because some of the ideas are so far as yet from being familiar that a certain amount of iteration may be harmless if not useful.

Clifford's actual intention with respect to all these writings was not to republish them as they stand, but to recast them in a book to be called 'The Creed of Science.' He had written in a note-book the following sketch of contents :—

THE CREED OF SCIENCE.

I. What ought we to believe?
1. The duty of inquiry and the sin of credulity.
2. The weight of authority.
3. The nature of inference.
    Is the order of the universe exact?
4. Is the order reasonable?

II. What is Science?
1. Conceptions and beliefs.
2. Knowledge is the guide of action.
3. My knowledge and our knowledge, or what is truth?
4. Truth for its own sake.

III. The History of the Sun.
1. The Sun's present work.
2. The evolution of the Earth's crust and evolution of life.
3. The age of the earth.
4. The formation of the solar system.

IV. Atoms.
1. The molecular hypothesis.
2. How far we know that it is true.
3. What we do not know.
4. The nature of the evidence for a [? the] hypothesis.

V. Ether.

    1. Light is a change of state periodic in time and space.

    2. Radiant heat (same thing) has energy, and therefore is motion of matter.

    3. Whatever motion is periodically reversed in light is continuous round an electric current.

    4. Difficulties.

VI. The beginning and the end.

    1. Are molecules eternal ?

    2. Thomson's hypothesis.

    3. The argument from dissipation.

    4. The limits of knowledge.

VII. Body and Mind.

    1. The atomism of the nervous system.

    2. The atomism of mind.

    3. The parallelism of the two.

    4. The great gulf fixed between them.

VIII. The Unseen Reality.

    1. There is no matter without something like mind behind it.

    2. All matter is a part of our minds.

    3. The material universe is a picture of something which is like mind.

    4. How far is it a true picture ?

IX. God and the Soul.

    1. Will and intelligence imply a certain organization of matter.

    2. No will or intelligence except those of men and animals has worked in the Solar System.

    3. The consciousness of man breaks up at the same time with his brain.

    4. Nature is uniform in human action.

X. Right and Wrong.

    1. The facts of the moral sense.

    2. The theory of responsibility.

    3. The foundation of absolute morality.

    4. Piety and Truth.

LECTURES AND ESSAYS

ON SOME OF THE CONDITIONS OF MENTAL DEVELOPMENT.[1]

IF you will carefully consider what it is that you have done most often during this day, I think you can hardly avoid being drawn to this conclusion : that you have really done nothing else from morning to night but *change your mind.* You began by waking up. Now that act of waking is itself a passage of the mind from an unconscious to a conscious state, which is about the greatest change that the mind can undergo. Your first idea upon waking was probably that you were going to rest for some time longer; but this rapidly passed away, and was changed into a desire for action, which again transformed itself into volition, and produced the physical act of getting up. From this arose a series of new sensations; that is to say, a change of mind from the state of not perceiving or feeling these things to the state of feeling them. And so afterwards. Did you perform any deliberate action? There was the change of mind from indecision to decision, from decided desire to volition, from volition to act. Did you perform an impulsive action? Here there is the more sudden and conspicuous change marked by the

[1] Discourse delivered at the Royal Institution, March 6, 1868.

word *impulsive*; as if your mind were a shuttlecock, which has its entire state of motion suddenly changed by the *impulse* of the battledore: conceive the shuttlecock descending quite regularly with a gentle corkscrew motion—the battledore intervenes—instantaneously the shuttlecock flies off in a totally unexpected direction, having apparently no relation to its previous motion ; and you will see how very apt and expressive a simile you use when you speak of certain people as having an *impulsive temperament*. Have you felt happy or miserable? It was a change in your way of looking at things in general; a transition, as Spinoza says, from a lower to a higher state of perfection, or *vice versâ*. In a word, whatever you have done, or felt, or thought, you will find upon reflection that you could not possibly be conscious of anything else than a change of mind.

But then, you will be inclined to say, this change is only a small thing after all. It does not penetrate beyond the surface of the mind, so to speak. Your character, the general attitude which you take up with regard to circumstances outside, remains the same throughout the day: even for great numbers of days. You can distinguish between individual people to such an extent that you have a general idea of how a given person will act when placed in given circumstances. Now for this to be the case, it is clear that each person must have retained his individual character for a considerable period, so as to enable you to take note of his behaviour in different cases, to frame some sort of general rules about it, and from them to calculate what he would do in any supposed given case. But is it true that this character or mark by which you know one person from another is absolutely fixed and unvarying? Do you not speak of the character of a child growing into that of a man : of a man in new circumstances being quite a different person from what he was before? Is it not regarded as the greatest stroke of art in a novelist that he should be able not merely to draw a character at any given time, but also to sketch the growth of it through the changing circumstances of life? In fact, if you consider a little further, you will see that it is not even true that a character remains the same for a single day : every circumstance, however trivial, that in any way affects the mind, leaves its mark, infinitely small it may be, imperceptible in itself, but yet more indelible than the stone-carved hieroglyphics of Egypt. And the sum of all these marks is precisely what we call the character, which is thus itself a history of the entire previous life of the individual ; which is therefore continually being added to, continually growing, continually in a state of change.

Let me illustrate this relation by the example of the motion of a planet. People knew, ages and ages ago, that a planet was a thing constantly moving about from one place to another; and they made continual attempts to discover the *character* of its motion, so that by observing the general way in which it went on, they might be able to tell where it would be at any particular time. And they invented most ingenious and complicated ways of expressing this character :

'Cycle on epicycle, orb on orb,'

till a certain very profane king of Portugal, who was

learning astronomy, said that if *he* had been present at
the making of the Solar System, he would have tendered
some good advice. But the fact was that they were all
wrong, and the real case was by no means so complicated as they supposed it to be. Kepler was the first
to discover what was the real character of a planetary
orbit; and he did this in the case of the planet Mars.
He found that this planet moved in an ellipse or oval
curve round the sun which was situated rather askew
near the middle. But upon further observation, this
was found to be not quite exact; the orbit itself is
revolving slowly round the sun, it is getting elongated
and then flattened in turns, and even the plane in which
the motion takes place sways slowly from side to side
of its mean position. Thus you see that although the
elliptic character of the motion does represent it with
considerable exactness for a long time together, yet this
character itself must be regarded as incessantly in a
state of gradual change. But the great point of the
comparison—to aid in the conception of which, in fact,
I have used the comparison at all—is this: that for no
two seconds together does any possible ellipse *accurately*
represent the orbit. It is impossible for the planet to
move a single inch on its way, without the oval having
slightly turned round, become slightly elongated or
shortened, and swayed slightly out of its plane; so that
the oval which accurately represented the motion at
one end of the inch would not accurately represent the
motion at the other end. The application is obvious.
In like manner it is true that the character which will
roughly represent the law of a man's actions for some
considerable time, will not *accurately* represent that

law for two seconds together. No action can take place in accordance with the character without modifying the character itself; just as no motion of a planet could take place along its orbit without a simultaneous change in the orbit itself.

But I will go even further. Historians are accustomed to say that at any given point of a nation's history there is a certain general type which prevails among the various changes of character which different men undergo. There is some kind of law, they say, which regulates the slow growth of each character from childhood to age; so that if you compared together all the biographies you would find a sort of family likeness suggesting that some common force had acted upon them all to make these changes. This force they call the Spirit of the Age. The spirit, then, which determines all the changes of character that take place, which is, therefore, more persistent than character itself,— is this, at last, a thing absolutely fixed, permanent, free from fluctuations? No: for the entire history of humanity is an account of its continual changes. It tells how there were great waves of change which spread from country to country, and swept over whole continents, and passed away; to be succeeded by similar waves. No history can be philosophical which does not trace the origin and course of these: things far more important than all the kings and rulers and battles and dates which some people imagine to be history.

To recapitulate. The mind is changing so constantly that we only know it by its changes. The law of these changes, which we call character, is also a thing which is continually changing, though more slowly. And that law of force which governs all the changes of character in a given people at a given time, which we call the Spirit of the Age, this also changes, though more slowly still.

Now it is a belief which, whether true or not, we are all of us constantly acting upon, that these changes have some kind of fixed relation to the surrounding circumstances. In every part of our conduct towards other people we proceed constantly upon the assumption that what they will do is to a certain extent, and in some way or other, dependent upon what we do. If I want a man to treat me with kindness and respect, I have to behave in a certain way towards him. If I want to produce a more special and defined effect, I have recourse to threats or promises. And even if I want to produce a certain change of mind in myself, I proceed upon the same assumption that in some way or other, and to a certain extent, I am dependent on the surrounding circumstances. People tie knots in their handkerchiefs to make themselves remember things; they also read definite books with a view of putting themselves into definite mental states or moods; and attempts are constantly made to produce even a further and more permanent effect, to effect an alteration in character. What else is the meaning of schools, prisons, reformatories, and the like? Some have actually gone further than this: there have not been wanting enterprising and far-seeing statesmen who have attempted to control and direct the Spirit of the Age. Now in all these cases in which we use means to an end, we are clearly proceeding on the assumption that there is some fixed relation of cause and effect, in virtue of which the means we adopt may be antecedently expected to bring about the end we are in pursuit of. We are all along assuming, in fact, that changes of mind are connected by some fixed laws or relations with surrounding circumstances. Now this being so, since every mind is thus continually changing its character for better or worse, and since the character of a race or nation is subject to the same constant change; since also these changes are connected in some definite manner with surrounding circumstances; the question naturally presents itself, What is that attitude of mind which is likely to change for the better? All the individuals of a race are changing in character, all changing in different · directions, with every possible degree of divergence; also the average character itself, the Spirit of the Age, is either changing in some one definite direction, or tending to split into two different characters: an individual, therefore, may be going with the race or dropping out of it; a portion of the race may be going right or wrong. Let us suppose that some portion of the race is going right and improving: the question is, In what way are we to distinguish that individual who is improving with the race, from the others who are either dropping out of the march altogether or going wrong?

Now what I have proposed to myself to do to-night is this, merely to suggest a method by which this question may ultimately be answered. I shall also endeavour afterwards to point out what I conceive to be one or two results of this method: but this part will be of minor importance; the results depend upon my

application of the method, can be only partially true,
and may be wholly false ; the method itself I believe to
be altogether a true one, and one which must ultimately
lead to the correct results.

It consists in observing and making use of a certain
analogy, namely, the analogy between the mind and the
visible forms of organic life. You know that every
animal and every plant is constantly going through a
series of changes. The flower closes at night and opens
in the morning ; trees are bare in winter and covered
with leaves in summer ; while the growth of every
organism from birth to maturity cannot fail to strike
you as a forcible illustration of the gradual change of
character in the human mind. In fact, it is the peculiarity of living things not merely that they change
under the influence of surrounding circumstances, but
that any change which takes place in them is not lost
but retained, and, as it were, built into the organism
to serve as the foundation for future actions. If you
cause any distortion in the growth of a tree and make
it crooked, whatever you may do afterwards to make
the tree straight, the mark of your distortion is there ;
it is absolutely indelible ; it has become part of the
tree's nature, and will even be transmitted in some small
degree to the seeds. Suppose, however, that you take a
piece of inanimate matter—a lump of gold, say, which
is yellow and quite hard—you melt it, and it becomes
liquid and green. Here an enormous change has been
produced ; but let it cool ; it returns to the solid and
yellow condition, and looks precisely as before—there
is no trace whatever of the actions that have been
going on. No one can tell by examining a piece of

gold how often it has been melted and cooled in geologic ages by changes of the earth's crust, or even in the last year by the hand of man. Anyone who cuts down an oak can tell by the rings in its trunk how many times winter has frozen it into widowhood and summer has warmed it into life. A living being must always contain within itself the history not merely of its own existence but of all its ancestors. Seeing then that in its continual changes and in the preservation of the records of those changes every organism resembles the mind, so that to this extent they belong to the same order of phenomena, may we not reasonably suppose that the laws of change are alike, if not identical, in the two cases? This is of course a mere supposition, not deducible from anything which we have yet observed, which requires therefore to be tested by facts. I shall endeavour to show that the supposition is well founded; that such laws of change as have been observed in animals and plants do equally hold good in the case of the mind. I shall then endeavour to find out what we mean by higher and lower in the two cases, and to show, in fact, that we mean much the same thing. Supposing all this to have been done, the question will have been stated in a form which it is possible to answer. I shall then make an attempt to give part of the answer to it.

In investigating the laws of change of organic beings I shall make use of what is called the Evolution-hypothesis, which, as applied to this subject, is much the same thing as the Darwinian theory, though it is not by any means tied down to the special views of Mr. Darwin. But I shall use this merely as an hypothesis; and the validity of the method of investigation which I have suggested is entirely independent of the truth of that hypothesis. If you will pardon me for a short time, I should like to illustrate somewhat further what I mean by this.

When Kepler found out what was the form of the orbit described by the planet Mars, he thought that the planet was driven by some force which acted in the direction in which the planet was going. I have known people who learned a certain amount of astronomy for nautical purposes, whose ideas were very similar to those of Kepler. They thought that the sun's rotation was what caused the planets to revolve about him, just as if you spin a teaspoon in the middle of a cup of tea, it makes the bubbles go round and round. But Newton discovered that the real state of the case was far different. If you fasten a ball on to the end of an elastic string, and then swing it round and round, you can make the ball describe an orbit very similar to that of the planet, so that your hand is not quite in the centre of it. Now here the pulling force does not act in the direction in which the ball is going, but always in the direction of your hand, and yet the ball revolves about your hand and never actually comes to it. Newton supposed that the case of the planet was similar to that of the ball; that it was always pulled in the direction of the sun, and that this attraction or pulling of the sun produced the revolution of the planet, in the same way that the traction or pulling of the elastic string produces the revolution of the ball. *What* there is between the sun and the planet that makes each of them pull the other, Newton did not know; nobody knows to this day; and all we are now able to assert positively is that the known motion of the planet is precisely what would be produced if it were fastened to the sun by an elastic string, having a certain law of elasticity. Now observe the nature of this discovery, the greatest in its consequences that has ever yet been made in physical science :—

I. It begins with an hypothesis, by supposing that there is an analogy between the motion of a planet and the motion of a ball at the end of a string.

II. Science becomes independent of the hypothesis, for we merely use it to investigate the properties of the motion, and do not trouble ourselves further about the cause of it.

I will take another example. It has been supposed for a long time that light consists of waves transmitted through an extremely thin ethereal jelly that pervades all space; it is easy to see the very rapid tremor which spreads through a jelly when you strike it at one point. From this hypothesis we can deduce laws of the propagation of light, and of the way in which different rays interfere with one another, and the laws so deduced are abundantly confirmed by experiment. But here also science kicks down the ladder by which she has risen. In order to explain the phenomena of light it is not necessary to assume anything more than a periodical oscillation between two states at any given point of space. *What* the two states are nobody knows; and the only thing we can assert with any degree of probability is that they are *not* states of merely mechanical displacement like the tremor of a jelly; for the phenomena of fluorescence appear to negative this supposition. Here again, then, the same two remarks may be made. The scientific discovery appears first as the hypothesis of an analogy; and science tends to become independent of the hypothesis.

The theory of heat is another example. If you hold one end of a poker in the fire, the other end becomes hot, even though it is not exposed to the rays of the fire. Fourier, in trying to find the laws of this spread of heat from one part of a body to another part, made the hypothesis that heat was a fluid which flowed from the hot end into the cold as water flows through a pipe. From this hypothesis the laws of conduction were deduced; but in the process it was found that the very same laws would flow from other hypotheses. In fact, whatever can be explained by the motion of a fluid can be equally well explained either by the attraction of particles or by the strains of a solid substance; the very same mathematical calculations result from the three distinct hypotheses; and science, though completely independent of all three, may yet choose one of them as serving to link together different trains of physical inquiry.

Now the same two remarks which may be made in all these cases apply equally to the evolution-hypothesis. It is grounded on a supposed analogy between the growth of a species and the growth of an individual. It supposes, for instance, that the race of crabs has gone through much the same sort of changes as every crab goes through now, in the course of its formation in the egg; changes represented by its pristine shape utterly unlike what it afterwards attains, and by its gradual metamorphosis and formation of shell and claws. By this analogy the laws of change are suggested, and these are afterwards checked and corrected by the facts. But as before, science tends to become independent of hypothesis. The laws of change are established for present and finitely distant times; but they give us no positive information about the origin of things. So, therefore, if I make use of this hypothesis to represent to you the laws of change that are deduced from it, you will see that the truth of those laws and the conclusions which may be drawn from them are in no way dependent on the truth of the hypothesis.

There are certain errors current about the nature of the evolution-theory which I wish particularly to guard against. In the first place it is very commonly supposed that all existing animals can be arranged in one continuous chain, from the highest to the lowest; that the transition is gradual all through, and that nature makes no jumps. This idea was worked out into a system of classification by Linnæus, and survived among naturalists until the time of Cuvier. ' They were bent,' says Agassiz, ' upon establishing one continual uniform series to embrace all animals, between the links of which it was supposed there were no unequal intervals.' . . . ' They called their system *la chaîne des êtres.*' The holders of the Darwinian theory are then supposed to believe that all these forms grew out of one another, beginning with the lowest and ending with the highest; so that any one animal of the series has in the course of its evolution passed through all the lower forms. And as the species is thus supposed to have grown up through the chain, and the lower species to be continually growing into the higher, so it is imagined that every individual creature, in the course of its production, passes through the lower adult forms; that a chicken, for instance, while it is being formed in the egg, becomes in succession a snail, an insect, a fish, and a reptile, before it becomes a bird. Now that all these ideas are entirely wrong, I need

hardly remind you; and I have mentioned them in order that there may be no mistake about the theory which I am using as an analogy. So far is it from being possible to arrange existing organisms in a single

line or chain, that they cannot be adequately represented even in the manner which is attempted in the
preceding diagram taken from Spencer's 'Principles of
Biology,' vol. i. p. 303.

In the next place, no existing organism could possibly grow into any other. What is really supposed is
this :—that if you went back a million years or so, and
made a picture like this one, representing the forms
that existed then, no single spot which is covered in
one figure would be covered in the other ; but the
general arrangement would be very similar, except
that all the groups would be nearer to the centre or
radiant point, and therefore nearer to each other.
And if you made a third picture, representing the
state of things another million years or so further
back, then they would be still nearer together ; and at
a distance of time too vast to be represented, they
would all converge into this radiant point. So the
theory is that at that stupendous distance of time all
species were alike, mere specks of jelly ; that they
gradually diverged from each other and got more and
more different, till at last they attained the almost infinite variety that we now find. If you will imagine a
tree with spreading branches, like an oak ; then the
outside leaves at any time may be taken to represent
all the existing species at a given time. It is quite impossible to arrange them in any serial order. As the
tree grows, the outer leaves diverge, and get further
from the trunk and from each other ; and two extremi-
ties that have once diverged never converge and grow
together again. But even this simile is insufficient ; for
species may diverge in a far greater variety of direc-

tions than the branches of a tree. Space has not dimensions enough to represent the true state of the case.

Von Baer's doctrine of development is illustrated by the same figure. If you took embryos of polypes, and snails, and cuttle-fish, and insects, and crabs, and fish, and frogs, and if you could watch their gradual growth into these several animals : at first they would be all absolutely alike and indistinguishable. Then, after a little while, you would find that they might be sorted off into these four great classes. Afterwards these groups might be divided into smaller groups, representing orders; then these into families and genera; last of all would appear those differences which would separate them into species.

The evolution-hypothesis, then, represents a *race* of animals or plants as a thing slowly changing : and it also represents these changes as connected by fixed laws with the action of the surrounding circumstances, or, as it is customary to say, the environment. Now the action of the environment on a race is of two kinds, direct and indirect. That part which is called direct action is very easily understood. There is no difficulty in seeing how changes of climate might produce changes in the colour of the skin, or how new conditions which necessitated the greater use of any organ would lead to the increase of that organ, as we know that muscles may be made to swell with exercise ; and changes thus made habitual would in time be inherited. But the indirect action of the environment, which is called natural selection, is still more important. The mode of its operation may be seen from an example. There are two butterflies in South America, nearly resembling one another in form, but one of which has a very sweet taste and is liked by the birds, while the other is bitter and distasteful to them. Now suppose that, for some reason or other, sweet butterflies were occasionally produced with markings similar to the bitter ones, these, being mistaken by the birds for bitter ones, would run less chance of being eaten, and therefore more chance of surviving and leaving offspring. If this peculiarity of marking is at all inheritable, then the number of sweet butterflies with bitter marks will in the next generation be greater in proportion to the whole number than before; and, as this process goes on, the sweet butter-flies which retain their distinguishing marks will be all weeded out by the birds, and the entire species will have copied the markings of the bitter species. This has actually taken place: the one species has mimicked the markings of the other. Here we see the working of Natural Selection. Any variation in an individual which gives him an advantage in the struggle for life is more likely to be transmitted to offspring than any other variation, because the individual is more likely to sur-vive; so that nature gradually weeds out all those forms which are not suited to the environment, and thus tends to produce equilibrium between the species and its surrounding circumstances. Changes, then, are pro-duced in a species by the selection of advantageous changes which happen to be made in individuals. Now there are three kinds of change that are produced in individuals: change of size, or growth; change of structure, that is to say, change in the shape and arrangement of the parts, as when the cartilaginous skeleton of an infant becomes hardened into bone; and change of function, that is to say, change in the use which is made of any part of the organism. I have one or two remarks to make about the first of these, namely, growth, or change of size. Every organism is continually taking in matter through the external surface to feed the inside. A certain quantity of this is needed to make up for the waste that is continually going on. But let us suppose, to begin with, that an organism has more surface than it absolutely wants to make up for waste, then a certain portion of the assimilated matter, or food, will remain over, and the organism will increase in size. But, you say, if this is all that is meant by growth why does it not go on for ever? The explanation is very simple. I take this cube, which has six sides, each a square inch; let us suppose it to represent an animal, and imagine, to begin with, that two of the sides by themselves are capable of feeding the whole mass, then the nutrition taken in by the other four sides is left over, and the mass must increase in size. Imagine it now grown to twice the linear dimensions, that is to say, to a cube every side of which is two inches. The mass to be fed is now eight times what it was, while the surface is only four times as great; of the twenty-four square inches of surface sixteen are taken up with feeding the mass, while only eight, or one-third, are left to supply the materials for growth. Still there is an overplus, and the organism will grow. Let it now acquire three times its original height and breadth and thickness, the mass is twenty-seven times as great, and the surface only nine times: that is to say, while there are twenty-seven cubic inches to be fed, there are just fifty-four square inches to feed them. There is no longer any overplus; the organism will stop growing. And it is a general rule that, in any case, when a thing grows its mass increases much faster than its surface. However much, therefore, the feeding power of the surface may be in excess to begin with, the mass must inevitably catch it up, and the growth will stop.

Now the changes of an individual mind may be reduced to the same three types :—

Growth.

Change of structure.

Change of function.

First, then, what is the growth of the mind? It is the acquisition of new knowledge; not merely of that which is required to make up for our wonderful power of forgetting, for oblivion is really a far more marvellous thing than memory; but of a certain overplus which goes to increase the entire mass of our mental experiences. Now I do not know whether there is any race between surface and mass here as in the case of an organism; but it is certainly true that whereas in childhood the amount we forget is very little, and our powers of acquisition preponderate immensely over our powers of oblivion; as we grow up, the powers of oblivion gain rapidly upon the acquisitive ones, and finally catch them up; the growth ceases as soon as this balance is attained. So that in this first law, you see, there is an entire analogy between the two cases.

In the next place, the mind experiences changes of structure; that is to say, changes in the shape and arrangement of its parts. Ideas which were only feebly connected become aggregated into a close and compact whole. The ideas of several different qualities, for instance, which we never thought of as connected with each other, are brought together by the qualities being found to exist in the same object. In this way we form conceptions of things, which gradually get so compact that we cannot even in thought separate them into their component parts. Portions of our knowledge which we held as distinct are connected together by scientific theories; images which were scattered all about are bound up into living bundles by the artist, and so we find them re-arranged.

Lastly, changes of function take place. Everybody knows how the mental faculties open out and become visible as a child grows up. Men acquire faculties by practice. And without any conscious seeking, you must know how often we wake up as it were and find ourselves gifted with new powers. We have found evidence then of the existence of our three types of change,—growth, structure, and function.

The actions therefore which go on between the environment and the individual may be reduced to the same three types in the case of the mind as in the case of any visible organism. Being somewhat encouraged by this result, let us go back to our original question. What is that attitude of mind which is likely to change for the better? What is the meaning of *better*?

Although it is quite impossible to arrange all existing organisms in a serial chain, yet we certainly have a general notion of higher and lower. A bird we regard as higher than a fish, and a dog is higher than a snake, And if we return to our illustration of the tree, we shall see that at every point, at any given time, there is a definite direction of development. So that though we might not be able to say which of two co-existing organisms was the higher, yet, by comparing a species with itself at a slightly later time, we might say whether it had degenerated or improved. Now by examining various cases, we shall find that there are six marks of improvement :—

The parts of the organism get more different.

The parts of the organism get more connected.

The organism gets more different from the environment.

The organism gets more connected with the environment.

The organism gets more different from other individuals.

The organism gets more connected with other individuals.

The processes in fact which result in development are made up of *differentiation* and *integration*; differentiation means the making things to be different, integration means the binding them together into a whole; these are applied to the parts of the organism, the organism and surrounding nature, the organism and other organisms. Differentiation of parts is illustrated by the figure on the next page. [Spencer's ' Principles of Biology,' vol ii. p. 187.]

Integration of parts means the connected play of them; so that one being affected the rest are affected. Differentiation from the environment takes place in weight, composition, and temperature. A polype is little else than sea-water, which it inhabits; a fish is several degrees of temperature above it, and made of quite different materials; till at last a mammal is 70° or 80° above the surrounding matter, and made of still more different materials. Integration with the environment means close correspondence with it; actions of the environment are followed by corresponding actions

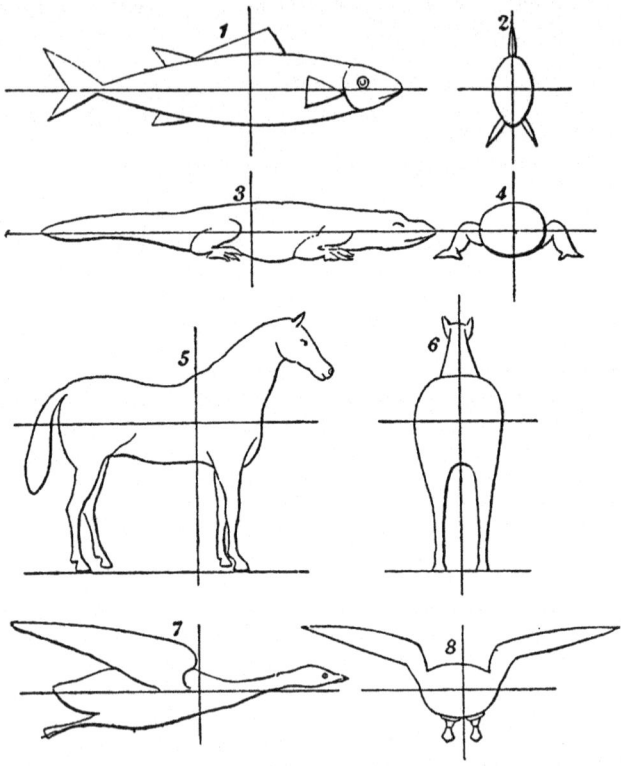

of the animal. Differentiation from other organisms means individuality; integration with them sociality.

In a similar way we have a sort of general notion of higher and lower stages of mental development. I will endeavour to show that this general notion resolves itself into a measure of the extent to which the same six processes have gone on, namely :—

Separation of parts,
Connexion of parts,
Separation from the environment,
Closer correspondence with the environment,
Separation from other individuals,
Sociality.

The only conception we can form of a purely unconscious state is one in which all is exactly alike, or rather, in which there is no difference.

> There is not one thing with another,
> But Evil saith to Good: My brother,
>   My brother, I am one with thee:
> They shall not strive nor cry for ever:
> No man shall choose between them: never
>   Shall this thing end and that thing be.

The first indication of consciousness is a perception of difference. The child's eyes follow the light. Immediately this colourless, homogeneous universe splits up into two parts, the light part and the dark part. A line is drawn across it, it is made heterogeneous, and the first thing that exists is a distinction. Then other lines are drawn; appearance is separated into white, black, blue, red, and so on. This is the first process, the differentiation of the parts of consciousness. But by-and-by a number of these lines of distinction are found to enclose a definite space; they assume relations to one another; the lines white, round, light, capable of being thrown at people, include the conception of a ball; this gains coherence, becomes one, a thing, holding itself together not only separated from the rest of consciousness, but connected in itself into a distinct whole, integrated. Here we have the second process. And throughout

our lives the same two processes go hand in hand ;
whatever we perceive is a line of demarcation between
two different things ; we can be conscious of nothing
but a separation, a change in passing from one thing
to another. And these different lines of demarcation
are constantly connecting themselves together, marking
out portions of our consciousness as complete wholes,
and making them cohere. Just as a sculptor clears
away from a block of marble now this piece and now
that, making every time a separation between what
is to be kept and what is to be chipped off, till at
last all these chippings manifest the connexion that
ran through them, and the finished statue stands out
as a complete whole, a positive thing made up of con-
tradictory negations : so is a conception formed in the
mind.

And this conception, when it is thus made into a
whole, integrated, by an act of the mind, what does it
immediately appear to be ? Why, something outside
of ourselves, a real thing, different from us. This is the
third process, the process of *differentiation* from the
environment. This is beautifully described by Cuvier,
who pictures the first man wandering about in ecstasies
at the discovery of so many new parts of himself; till
gradually he learns that they are not himself, but
things outside. This notion, then, of a thing being
real, existing external to ourselves, is due to the active
power of the mind which regards it as one, which binds
together all its boundaries. And this goes on as long
as we live. Constantly we frame to ourselves more
complicated combinations of ideas, and by giving them
unity make them real. And, at the same time, the

converse process is equally active. While more and
more of our ideas are put outside of us and made real,
our minds are continually growing more and more into
accordance with the nature of external things; our
ideas become truer, more conformable to the facts; and
at the same time they answer more surely and completely to changes in the environment; a new experience is more rapidly and more completely connected
with the sum of previous experiences. But there is
more than this. The action of these two laws taken
together does in fact amount to the creation of new
senses. Men of science, for example, have to deal
with extremely abstract and general conceptions. By
constant use and familiarity, these, and the relations
between them, become just as real and external as the
ordinary objects of experience; and the perception of
new relations among them is so rapid, the correspondence
of the mind to external circumstances so great, that a
real scientific sense is developed, by which things are
perceived as immediately and truly as I see you now.
Poets and painters and musicians also are so accustomed to put outside of them the idea of beauty, that
it becomes a real external existence, a thing which they
see with spiritual eyes, and then describe to you, but
by no means create, any more than we seem to create
these ideas of table and forms and light, which we put
together long ago. There is no scientific discoverer,
no poet, no painter, no musician, who will not tell you
that he found ready-made his discovery, or poem or
picture—that it came to him from outside, and that he
did not consciously create it from within. And there
is reason to think that these senses or insights are

things which actually increase among mankind. It is
certain, at least, that the scientific sense is immensely
more developed now than it was three hundred years
ago ; and though it may be impossible to find any
absolute standard of art, yet it is acknowledged that a
number of minds which are subject to artistic training
will tend to arrange themselves under certain great
groups, and that the members of each group will give
an independent yet consentient testimony about artistic
questions.   And this arrangement into schools, and the
definiteness of the conclusions reached in each, are on
the increase, so that here, it would seem, are actually
two new senses, the scientific and the artistic, which the
mind is now in the process of forming for itself.   There
are two remaining marks of development : differentiation from surrounding minds, which is the growth of
individuality ; and closer correspondence with them,
wider sympathies, more perfect understanding of others.
These, you will instantly admit, are precisely the twin
characteristics of a man of genius.   He is clearly
distinct from the people that surround him, that is how
you recognize him ; but then this very distinction must
be such as to bind him still closer to them, extend and
intensify his sympathies, make him want their wants,
rejoice over their joys, be cast down by their sorrows.
Just as the throat is a complicated thing, quite different
from the rest of the body, but yet is always ready to
cry when any other part is hurt.

　　We have thus got a tolerably definite notion of
what mental development means.   It is a process of
simultaneous differentiation and integration which goes
on in the parts of consciousness, between the mind and

external things, between the mind and other minds. And the question I want answered is, What attitude of mind tends to further these processes?

I have now done all that it was my business to do, namely, I have stated the question in a form in which it is possible to answer it. There is no doubt that by a careful study of the operations of nature we shall be able to find out what actions of an organism are favourable to its higher development. Having formulated these into a law, we shall be able to interpret this law with reference to the mind.

But now I am going to venture on a partial answer to this question. What I am going to say is mere speculation, and requires to be verified by facts.

The changes which take place in an organism are of two kinds. Some are produced by the direct action of things outside, and these are to a great extent similar to the changes which we observe in inanimate things. When a tree is bent over by the wind and gets ultimately fixed in this position, the change is in no way different from that which takes place when we bend a wire and it does not entirely return to its former straightness. Other changes are produced by the spontaneous action of that store of force which by the process of growth is necessarily accumulated within the organism. Such are all those apparently disconnected motions which make up the great distinction between living things and dead. Now my speculation is, that advantageous permanent changes are always produced by the spontaneous action of the organism, and not by the direct action of the environment. This,

I think, is most clear when we take an extreme case. Let us suppose a race of animals that never had any changes produced by their spontaneous activity. The race must at a certain time have a definite amount of plasticity, that is, a definite power of adapting itself to altered circumstances by changing in accordance with them. Every permanent effect of the environment upon them is a crystallization of some part which before was plastic; for the part must have been plastic for the effect to be produced at all; and as the effect is permanent, the part has to that extent lost in plasticity. As this goes on, the race of animals will bind up in itself more and more of its history, but will in that process lose the capability of change which it once had; at last it will be quite fixed, crystallized, incapable of change. Then it must inevitably die out in time; for the environment must change sooner or later, and then the race, incapable of changing in accordance with it, must be killed off. On the other hand, any addition to the organism which is made by its spontaneous activity is an addition of something which has not yet been acted upon by the environment, which is therefore plastic, capable of indefinite modification, in fact, an increase of power. The bending of a tree by the wind is a positive disadvantage to it if the wind should ever happen to blow from the other side. But when a plant, for no apparent reason, grows long hairs to its seed—the material for which may have been accidentally supplied by the environment, while its use in this way is a spontaneous action of the plant—this is a definite increase of power; for the new organ may be modified in any conceivable way to suit the exigencies of the environment, may cling to the sides of beasts, and so help the distribution of the seed, or effect the same object by being caught by the wind. Activity, in fact, is the first condition of development. A very good example of this occurs in Professor Huxley's lizards, of which you heard two or three weeks ago.[1] About the time marked by the Primary strata it appears that there was a race of lizards, thirty feet high, that walked on their hind legs, balancing themselves by their long tails, and having three toes like birds. This race diverged in three directions. Some of them yielded to the immediate promptings of the environment, found it convenient to go on all fours and eat fish ; they became crocodiles. Others took to exercising their fore-legs violently, developed three long fingers, and became birds. The rest were for a long while undecided whether they would use their arms or their legs most ; at length they diverged, and some became pterodactyles and others kangaroos. For Mr. Seeley, of Cambridge, has discovered marsupial bones in pterodactyles ; that is to say, bones like those which were supposed peculiar to the order of mammals to which the kangaroo belongs.

Assuming now that this law is true, and that the development of an organism proceeds from its activities rather than its passivities, let us apply it to the mind. What, in fact, are the conditions which must be satisfied by a mind in process of upward development, so far as this law gives them ?

They are two ; one positive, the other negative.

[1] ['On the animals which are most nearly intermediate between birds and reptiles,' *Roy. Inst. Proc. V.* 1869, p. 278.]

The positive condition is that the mind should act
rather than assimilate, that its attitude should be one
of creation rather than of acquisition. If scientific, it
must not rest in the contemplation of existing theories,
or the learning of facts by rote; it must act, create,
make fresh powers, discover new facts and laws. And,
if the analogy is true, it must create things not immediately useful. I am here putting in a word for those
abstruse mathematical researches which are so often
abused for having no obvious physical application.
The fact is that the most useful parts of science have
been investigated for the sake of truth, and not for their
usefulness. A new branch of mathematics, which has
sprung up in the last twenty years, was denounced by
the Astronomer Royal before the University of Cambridge as doomed to be forgotten, on account of its
uselessness. Now it turns out that the reason why we
cannot go further in our investigations of molecular
action is that we do not know enough of this branch
of mathematics. If the mind is artistic, it must not sit
down in hopeless awe before the monuments of the
great masters, as if heights so lofty could have no
heaven beyond them. Still less must it tremble before
the conventionalism of one age, when its mission may
be to form the whole life of the age succeeding. No
amount of erudition or technical skill or critical power
can absolve the mind from the necessity of creating, if
it would grow. And the power of creation is not a
matter of static ability, so that one man absolutely can
do these things and another man absolutely cannot; it
is a matter of habits and desires. The results of things

follow not from their state but from their tendency. The first condition then of mental development is that the attitude of the mind should be creative rather than acquisitive: or, as it has been well said, that intellectual food should go to form mental muscle and not mental fat.

The negative condition is plasticity: the avoidance of all such crystallization as is immediately suggested by the environment. A mind that would grow must let no ideas become permanent except such as lead to action. Towards all others it must maintain an attitude of absolute receptivity; admitting all, being modified by all, but permanently biassed by none. To become crystallized, fixed in opinion and mode of thought, is to lose the great characteristic of life, by which it is distinguished from inanimate nature: the power of adapting itself to circumstances.

This is true even more of the race. There are nations in the East so enslaved by custom that they seem to have lost all power of change except the capability of being destroyed. Propriety, in fact, is the crystallization of a race. And if we consider that a race, in proportion as it is plastic and capable of change, may be regarded as young and vigorous, while a race which is fixed, persistent in form, unable to change, is as surely effete, worn out, in peril of extinction; we shall see, I think, the immense importance to a nation of checking the growth of conventionalities. It is quite possible for conventional rules of action and conventional habits of thought to get such power that progress is impossible, and the nation only fit to be im-

proved away. In the face of such a danger *it is not
right to be proper.*

NOTE.—*The following letter, published in the* 'Pall
Mall Gazette' *of June* 24, 1868, *should be read in connexion with this Discourse.*

Sir,—I ask for a portion of your space to say some-
thing about a lecture, ' On some of the Conditions of
Mental Development,' which I delivered at the Royal
Institution in March last.

In that lecture I attempted to state and partially
answer the question, ' What is that attitude of mind
which is most likely to change for the better?' I
proposed to do this by applying the hypothesis of the
variability of species to the present condition of the
human race. I put forward also for this purpose a
certain biological law, viz., that permanent advantageous
changes in an organism are due to its spontaneous
activity, and not to the direct action of the environ
ment.

In the short account of the evolution-hypothesis
which I prefixed, I followed Mr. Herbert Spencer's
' Principles of Biology,' not knowing, at the time, how
much of the theory was due to him personally, but
imagining that the greater part of it was the work of
previous biologists. On this account I omitted to make
such references to my special sources of information as
I should otherwise have made. I was also ignorant of
the developments and applications of the theory which
he has made in his other works, in which a great portion

of my remarks had been anticipated. These omissions I desire now to rectify.

Mr. Spencer's theory is to the ideas which preceded it even more than the theory of gravitation was to the guesses of Hooke and the facts of Kepler.

Finding only a vague notion of progress from lower to higher, he has affixed the specific meaning to the word *higher* of which I gave an account, defining the processes by which this progress is effected. He has, moreover, formed the conception of evolution as the subject of general propositions applicable to all natural processes, a conception which serves as the basis of a complete system of philosophy. In particular, he has applied this theory to the evolution of mind, developing the complete accordance between the laws of mental growth and of the growth of other organic functions. In fact, even if the two points which I put forward as my own—viz., the formal application of the biological method to a certain special problem, and the biological law which serves as a partial solution of it—have not before been explicitly developed (and of this I am not sure), yet they are consequences so immediate of the general theory that in any case the credit of them should entirely belong to the philosopher on whose domains I have unwittingly trespassed. The mistake, of course, affects me only, and could in no way injure the fame of one whose philosophical position is so high and so assured.

I may perhaps be excused for anticipating here what I hope to say more at length at another time,[1] that in

---

[1] This intention was never carried out, so far as the Editors are aware.

my belief the further deductions to be made from this theory, with reference to modern controversies, will lead to results at once more conservative, and in a certain sense more progressive, than is commonly supposed.

I remain, Sir, yours, &c.,

W. K. CLIFFORD.

ON THEORIES OF THE PHYSICAL FORCES.[1]

[REFERRING to the passage in ' Faust,'

> ' Geschrieben steht: Im Anfang war das Wort.
> Hier stock' ich schon! Wer hilft mir weiter fort?
> Ich kann das Wort so hoch unmöglich schätzen,
> Ich muss es anders übersetzen,
> Wenn ich vom Geiste recht erleuchtet bin.
> Geschrieben steht: Im Anfang war der Sinn.
> Bedenke wohl die erste Zeile,
> Dass deine Feder sich nicht übereile!
> Ist es der Sinn, der alles wirkt und schafft?
> Es sollte stehn: Im Anfang war die Kraft!
> Doch, auch indem ich dieses niederschreibe,
> Schon warnt mich was, dass ich dabei nicht bleibe.
> Mir hilft der Geist! Auf einmal seh' ich Rath,
> Und schreibe getrost: Im Anfang war die That!'

the speaker regarded it as a description of four views
or stages of opinion through which a man looking for
himself on the face of things is likely to pass; through
which also successive generations of the men who look
for themselves on the face of things are likely to pass.
He considered that by far the larger portion of scientific
thought at the present day is in the third stage—that,
namely, in which Force is regarded as the great fact

---

[1] Discourse delivered at the Royal Institution, February 18, 1870.
This discourse is reprinted as it stands in the Proceedings of the Royal
Institution. The opening paragraphs, being reported in the third person
and apparently abridged, are enclosed in square brackets.

that lies at the bottom of all things ; but that this is so far from being the final one, that even now the fourth stage is on its heels.   In the fourth stage the conception of Force disappears, and whatever happens is regarded as a deed.   The object of the discourse was to explain the nature of this transition, and to introduce certain conceptions which might serve to prepare the way for it.

There are, then, to be considered two different answers to the question, 'What is it that lies at the bottom of things ? '   The two answers correspond to two different ways of stating the question ; namely, first, 'Why do things happen ? ' and, secondly, 'What is it precisely that does happen ? '   The speaker maintained that the first question is external to the province of science altogether, and science has nothing to do with it ; but that the second is exactly the question to which science is always trying to find the answer.   It may be doubted whether the first question is within the province of human knowledge at all.   For it is as necessary that a question should *mean something*, in order to be a real question, as that an answer should mean something, in order to be a real answer.   And it is quite possible to put words together with a note of interrogation after them without asking any real question thereby.   Whether the phrase, 'Why do things happen ? ' as applied to physical phenomena, is a phrase of this kind or no, is not here to be considered. But that to the scientific enquirer there is not any '*why*' at all, and that if he ever uses the word it is always in the sense of *what*, the speaker regarded as certain.   In order to show what sort of way an exact knowledge of the facts would supersede the enquiry

after the cause of them, he then made use of the hypothesis of continuity ; showing, in the following manner,
that it involves such an interdependence of the facts
of the universe as forbids us to speak of one fact or set of
facts as the *cause* of another fact or set of facts.]

*The hypothesis of the continuity of space and time is
explained, and the alternative hypothesis is formulated.*

*From the hypothesis of the complete continuity of time-
changes, a knowledge of the entire history of a single particle is shown to be involved in a complete knowledge of
its state at any moment.*

Things frequently move. Some things move faster
than others. Even the same thing moves faster at one
time than it does at another time. When you say that
you are walking four miles an hour, you do not mean
that you actually walk exactly four miles in any
particular hour ; you mean that if anybody did walk
for an hour, keeping all the time exactly at the rate
at which you are walking, he would in that hour walk
four miles. But now suppose that you start walking
four miles an hour, and gradually quicken your pace,
until you are walking six miles an hour. Then this
question may be asked : Suppose that anybody chose a
particular number between four and six, say four and
five-eighths, is it perfectly certain that at some instant
or other during that interval you were walking at the
rate of four miles and five-eighths in the hour ? Or, to
put it more accurately, suppose that we have a vessel
containing four pints of water exactly, and that somebody adds to it a casual quantity of water less than two
pints. Then is it perfectly certain that between these
two times, when you were walking at four miles an

hour, and when you were walking six miles an hour, there was some particular instant at which you were walking exactly as many miles and fractions of a mile an hour as there are pints and fractions of a pint of water in the vessel? The hypothesis of continuity says that the answer to this question is yes; and this is the answer which everybody gives nowadays; which everybody has given mostly since the invention of the differential calculus.

But this is a question of fact, and not of calculation. Let us, therefore, try and imagine what the contrary hypothesis would be like.

You know what a 'wheel of life' is. There is a cylinder with slits in its side, which can be spun round rapidly; and you look through the slits at the pictures opposite. The result is that you see the pictures moving; moreover, you see them move faster or slower according as you turn the cylinder faster or slower. This is what you see, and what appears to happen; but now let us consider what actually does happen. I remember in particular a picture of a man rolling a ball down an inclined plane towards you; he was standing at the farther edge of the inclined plane, as it were behind a counter, and he picked up the balls one by one and rolled them towards you. But now when you took out the strips of paper on which the pictures were drawn, you found that they were really pictures of this man and his ball in a graduated series of positions. Each picture, of course, was perfectly still in itself, a mere drawing on the paper. The first one represented him with his hand below the counter; just picking up the ball; in the next, he had the ball in his hand, drawn back ready to roll down; in the next, the hand was thrown forwards with the ball in it; in the next, the ball had just left his hand and rolled a little way down; in the next farther, and so on. Now, these pictures being put in the inside of the cylinder which is turning round, come opposite you one by one. But you do not look directly at them; there are slits interposed. The effect of that is, that if you look straight at a certain portion of the opposite picture you can only see it for a very small interval of time; that, namely, during which the slit is passing in front of your eye. Now let us carefully examine what happens. When the slit passes, it goes so quickly that you get, as it were, almost an instantaneous photograph on your eye of the opposite picture; say of the man with his hand below the counter. Then this is effaced, and you see absolutely nothing until the next slit passes. But by the time the next slit comes, another picture has got opposite to you; so that you get an instantaneous photograph this time of the man with his hand drawn back and the ball in it. Then this in its turn is effaced, for a time you see nothing, and then you are given an instantaneous glimpse of the hand thrown forward. In this way, what you really see is darkness relieved by regularly-recurring glimpses of the figure in different positions. Now, this experience that you get is obviously consistent with the hypothesis that the man goes on moving all the time when he is hidden from you; so as to be in exactly that series of positions when you do catch a glimpse of him. And, in fact, you do instinctively, by an inevitable habit, admit this hypothesis, not merely into your mind as a speculation, but

into your very sensation as an observed phenomenon.
You simply see the man move; and, except for a
certain weariness in the eyes, there is nothing to distinguish this perception of movement from any other
perception of movement. At the same time we do
know very distinctly, and beyond the shadow of a
doubt, that there is no continuity in the picture at all:
that, in fact, you do not see the same picture twice
following, but a new one every time till the cycle is
completed; and that the picture never is in any position intermediate between two successive ones of those
which you see. Here then is an apparently continuous
motion which is really discontinuous; and moreover
there is an apparently continuous perception of it which
is really discontinuous—that is, it seems to be gradually
changed, while it really goes by little jumps.

I suppose very few people have looked at this toy
without wondering whether it is not actually and truly
a wheel of life, without any joke at all. I mean, that
it is very natural for the question to present itself, Do
I ever really see anything move? May not all my
apparently continuous perceptions be ultimately made
up of little jumps, which I run together by this same
inevitable instinct? There is another way in which
this is sometimes suggested. If you move your hand
quickly, you can see a continuous line of light, because
the image of every position of your hand lingers a little
while upon the retina. But now, if you do this in a
room lighted only by an electric spark which is not
going very fast, so that the general result is darkness
broken by nearly instantaneous flashes at regular intervals; then, instead of seeing a continuous line of light,

you will see a distinct series of different hands, perhaps about an inch apart, if the electric spark is going very slowly, and you move your hand very quickly. But now make the spark go quicker, or your hand slower; the distances between these several hands will gradually diminish, till—you do not know how—the continuous line of light is restored. And the question inevitably presents itself—is not every case of apparently continuous perception really a case of successive distinct images very close together?

That is to say, for instance, if I move my hand so in front of me, and apparently see it take up in succession every possible position on its path between the two extreme positions; do I really see this, or do I only see my hand in a certain very large number of distinct positions, and not at all in the intervening spaces?

I have no doubt whatever myself, that the latter alternative is the true one, and that the wheel of life is really an illustration and type of every moment of our existence. But I am not going to give my reasons for this opinion, because it is quite a different question from the one I am trying to get at. The question, namely, is this. What I see, or fancy I see, is quite consistent with the hypothesis that my hand really does go on moving continuously all the time, and takes up an infinite number of positions between the two extreme ones. But if this hypothesis is not true, what is true? and how are we to imagine any other state of things than that supposed by the hypothesis of continuity?

I draw here two rows of points. The upper row of points is to represent a series of positions in space which it is conceivable that a certain thing might take up.

The lower row of points is to represent a series of instants in time at which it is conceivable that the same thing might exist. Suppose now that at the instant of time represented by the first point of the lower row, the thing held the position in space represented by the first point of the upper row. Suppose that it only existed there for that instant, and then disappeared utterly, so that at these succeeding instants where the lower points have no points directly above them the thing is nowhere at all. Lastly, suppose that at this instant of time which *has* a space-dot above it, the thing existed in that space-position; and so on all through, the thing only existing at those instants whose representative points have a space-dot exactly above them, and being then in the space-position signified by such dots. Then we may call this a discontinuous motion; a motion because the thing is in different places at different times, though it is not at all times that it exists at all; and a discontinuous motion because the thing passes from one position to another *distant* from it without going through any intermediate position.

Now imagine that in each of these two series the dots are very close together indeed, and very great in number; so that, however small one made them on the paper, the lines would look as if they were continuous lines. And let the thing be a white speck travelling along the upper line in the manner I have described; namely, existing only when there is one dot exactly over another; only that as the lower dots represent instants of time, we may make some definite supposition and assume that one inch of them represents a second.

Then it is clear that if the dots were taken close enough together, and enough of them, the appearance would be precisely what we ordinarily see when a white speck moves along a line. That is to say, we have got some sort of representation of what we might have to suppose, if we did not assume the truth of the law of continuity.

You must here notice in particular that I suppose the series of positions denoted by the upper dots to be *all* the positions that are between the two end ones; that is, I suppose the path from one of these end ones to the other to be made up of a series of discrete positions. And similarly I suppose the series of instants denoted by the lower dots to be *all* the time that elapses between the two end ones; that is, I suppose the interval of time to contain a perfectly definite number of instants, these being further indivisible. Or we may say that on this alternative hypothesis space and time are discontinuous; that is, they are in separate parts which do not hold together. Now I must beg you to remember for a little while what the hypothesis of continuity is *not*, for I shall have to refer to this point again subsequently. In this kind of jumping motion that we have been imagining, the rate of motion of a thing could only be measured by the size of one of its jumps; that is, by the number of positions it passed over between two existences compared with the number of instants passed over. And this rate might obviously change by jumps as violent and sudden as those of the thing itself; at any instant when the thing was non-existent its rate would be non-existent, and whenever the thing came into existence its rate would suddenly have a value depending on how far off its last position was. In this case, therefore, our question about the intermediate rate—whether between walking four miles an hour and walking six miles an hour you must necessarily walk at all intermediate paces—must be answered in the negative. Now then, at last, let us investigate some consequences of supposing that motion is really continuous as it seems to be.

First, how to measure the rate at which a thing is moving? This was done experimentally by Galileo in the case of falling bodies, and I shall have to speak again of the results which he obtained. But at present I want to speak not of an experimental method of finding the rate, but of a theoretical method of representing it, invented by Newton, and called the curve of velocities.

Suppose that a point N is going along the line O Y, sometimes fast and sometimes slow; and that a point M

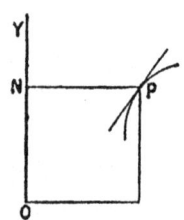

is going along the line O X always at the same rate. Also somebody always holds a stick N P so as to move with the point N, and be horizontal; and somebody holds a stick M P so as to move with the point M, and be vertical; and a third person keeps a pencil pressed in the corner where the two sticks cross at P. Then when the points M and N move, the point P will move too; and its motion will depend on that of the two other points. For instance, if the point N moves always exactly as fast as the point M, then the point P will go along the line O P midway between the lines O X O Y. If N moves twice as fast as M always, the point P will go along a line

nearer O Y ; and if N moves only half as fast as M, then
P will go along a line nearer O X. And in general, the
faster N moves, the more the line will be tilted
up ; and if the rate at which N goes is changeable, the direction of P's motion will be changeable, and P will then describe a curve, which
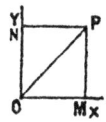
will be very steep when N is going fast, and more flat when
N is going slow. So that the steepness of this curve is
now a visible measure of the rate at which N is going,
and the *curvature* of it is a visible expression of the fact
that the rate is changeable. Now the hypothesis of
continuity in the motion of N asserts not merely that N
itself moves without any jumps, but that the rate at
which N is going changes gradually without any jumps,
and consequently that the direction of P's motion
changes gradually ; or that the curve described by P
cannot have a sharp point like this. But it asserts a
great deal more besides this, which I shall now endeavour to explain. Let us imagine a new point N_1, so
moving that whenever the old N is going at four
inches a second, N_1 shall be four inches from O ; and
when N is going at two inches a second, N_1 shall be two
inches from O, and so on, the distance of N_1 from O
being always exactly as far as N would go in a second
if it went at the rate at which it was moving at that
instant. Then the distance O N_1 measures the *rate* at
which N is going, or the *velocity* of N. If, for example,
there was a thing like a thermometer hung up in a
train, so that the height of the mercury always indicated
how fast the train was going ; when the train was going
17 miles an hour, the mercury stood at 17 inches, and
so on ; then the top of the mercury would behave to-

wards the train exactly as I want the point N_1 to behave towards the point N. It is to indicate by its height how fast N is going.

If, then, the velocity of N is changeable, the point N_1 will move up and down; and the rate at which N_1 moves up or down is clearly the rate at which the velocity of N is increasing or diminishing. This rate at which the velocity of N changes is called its acceleration. To return to our gauge inside of a train, if in the course of a minute it went up from 17 to 19, the train would be said to have an acceleration of two miles an hour per minute.

Now I shall take another point N_2, which is to behave towards N_1 exactly as N_1 behaves towards N; namely, the distance of N_2 from O_2 is to be always equal to the number of inches which N_1 is going in a second. And then I shall take a point N_3, related in just this same way to N_2, and so on, until I come to a point that does not move at all; and that I might never come to, so that I should have to go on taking new points for ever. But suppose now that I have got this series of points, and that they are all moving together. Then first of all there is my point N, which moves anyhow. Next there is N_1, such that $O_1 N_1$ is the velocity of N, or the rate of change of N's position. Next there is N_2, such that $O_2 N_2$ is the acceleration of N, or the rate of change of the rate of change of N's position. Then again $O_3 N_3$ is the change of the acceleration of N, or the rate of change of the rate of change of the rate of change of N's position, and so on. We may, if we like, agree to call the velocity of N the change of the first order, the velocity of N_1 the change of the second order, and so on.

Then the hypothesis of the perfect continuity of N's motion asserts that all these points move continuously without any jumps. Now, a jump made by any one of these points, being a finite change made in no time, would be a change made at an infinite rate; the next point, therefore, and all after it, would go right away from o, and disappear altogether. We may thus express the law of continuity also in this form; that there is no infinite change of any order.

Now, observe further that the rate at which anything is going is a property of the thing at that instant, and exists whether the thing goes any more or not. If I drop a marble on the floor, it goes faster and faster till it gets there, and then stops; but at the instant when it hit the floor it was going at a perfectly definite rate, which can be calculated, though it did not actually go any more.

In the same way the configuration of all these points which depend on the point N is a property of its motion at any given instant, quite independent of the continuance of that motion. I want you to take particular notice of this fact, that as the point N moves about, the whole set of points connected with it moves too; and that you may regard them as connected by some machine, which you may stop at any moment to contemplate the simultaneous positions of all these points; and that this set of simultaneous positions belongs just simply to that one position of the point N, and therefore to one instant of time.

Now I am going to state to you dogmatically a certain mathematical theorem, called Taylor's theorem; whereby you will see the very remarkable consequences of this hypothesis that we have made.

Namely, there is a certain rule whereby when the positions of all these points are known for any particular instant of time, then their positions at any other instant of time may be calculated from these; and it is impossible that they should have at that other instant any other positions than those so calculated. Provided always that there is no infinite change of any order; that is to say, that no one of the points has taken a sudden jump and sent all the points after it away to an infinite distance from o at any instant between the one for which the positions are given and the one for which they are calculated.

Remember that the positions of all the derivative points are mere properties of the motion of the point N at any instant; that in fact we must know them all in order to know completely the state of the point N at that instant. And then observe the result that we have arrived at. From the knowledge of the complete state at any instant of a thing whose motion obeys the law of continuity, we can calculate where it was at any past time, and where it will be at any future time. Now the hypothesis of continuity, of which we have only got disjointed fragments hitherto, is this; that the motion of every particle of the whole universe is entirely continuous. It follows from this hypothesis that the state at this moment of any detached fragment—say a particle of matter at the tip of my tongue—is an infallible record of the eternal past, an infallible prediction of the eternal future.

This is not the same as the statement that a complete knowledge of the position and velocity of every body in the universe at a given moment would suffice to determine the position at any previous or subsequent moment. That depends on an entirely different hypothesis, and relates to the whole, while this proposition that I am now expounding relates to every several part however small. Now reflect upon the fact that for a single particle—quite irrespective of everything else—the history of eternity is contained in every second of time ; and then try if you can find room in this one stifling eternal fact for any secondary causes and the question why? Why does the moon go round the earth? When the Solar system was nebulous, anybody who knew all about some one particle of nebulous vapour might have predicted that it would at this moment form part of the moon's mass, and be rotating about the earth exactly as it does. But why with an acceleration inversely as the square of the distance? There is no why; the fact is probably equivalent to saying that the continuous motion of one body is such as not to interfere with the continuous motion of another. If once so, then always ; the cause is only the fact that at some moment the thing is so,—or rather, the facts of one time are not the cause of the facts of another, but the facts of all time are included in one statement, and rigorously bound up together.

Parallel, however, with this hypothesis of temporal continuity, there is another hypothesis, not so universally held, of a continuity in space ; for which indeed I hope to make more room presently. And out of this it appears that as the history of eternity is written in every second of time, so the state of the universe is written in every point of space.

ON THE AIMS AND INSTRUMENTS OF SCIENTIFIC THOUGHT.[1]

It may have occurred (and very naturally too) to such as have had the curiosity to read the title of this lecture, that it must necessarily be a very dry and difficult subject; interesting to very few, intelligible to still fewer, and, above all, utterly incapable of adequate treatment within the limits of a discourse like this. It is quite true that a complete setting-forth of my subject would require a comprehensive treatise on logic, with incidental discussion of the main questions of metaphysics; that it would deal with ideas demanding close study for their apprehension, and investigations requiring a peculiar taste to relish them. It is not my intention now to present you with such a treatise.

The British Association, like the world in general, contains three classes of persons. In the first place, it contains scientific thinkers; that is to say, persons whose thoughts have very frequently the characters which I shall presently describe. Secondly it contains persons who are engaged in work upon what are called scientific subjects, but who in general do not, and are not expected to, think about these subjects in a scien-

[1] A Lecture delivered before the members of the British Association, at Brighton, on August 19, 1872.

tific manner. Lastly, it contains persons who suppose that their work and their thoughts are unscientific, but who would like to know something about the business of the other two classes aforesaid. Now, to anyone who belonging to one of these classes considers either of the other two, it will be apparent that there is a certain gulf between him and them; that he does not quite understand them, nor they him; and that an opportunity for sympathy and comradeship is lost through this want of understanding. It is this gulf that I desire to bridge over, to the best of my power. That the scientific thinker may consider his business in relation to the great life of mankind; that the noble army of practical workers may recognize their fellowship with the outer world, and the spirit which must guide both; that this so-called outer world may see in the work of science only the putting in evidence of all that is excellent in its own work,—may feel that the kingdom of science is within it: these are the objects of the present discourse. And they compel me to choose such portions of my vast subject as shall be intelligible to all, while they ought at least to command an interest universal, personal, and profound.

In the first place, then, what is meant by scientific thought? You may have heard some of it expressed in the various Sections this morning. You have probably also heard expressed in the same places a great deal of unscientific thought; notwithstanding that it was about mechanical energy, or about hydrocarbons, or about eocene deposits, or about malacopterygii. For scientific thought does not mean thought about scientific subjects with long names. There are no scientific subjects. The subject of science is the human universe; that is to say, everything that is, or has been, or may be related to man. Let us then, taking several topics in succession, endeavour to make out in what cases thought about them is scientific, and in what cases not.

Ancient astronomers observed that the relative motions of the sun and moon recurred all over again in the same order about every nineteen years. They were thus enabled to predict the time at which eclipses would take place. A calculator at one of our observatories can do a great deal more than this. Like them, he makes use of past experience to predict the future; but he knows of a great number of other cycles besides that one of the nineteen years, and takes account of all of them; and he can tell about the solar eclipse of six years hence exactly when it will be visible, and how much of the sun's surface will be covered at each place, and, to a second, at what time of day it will begin and finish there. This prediction involves technical skill of the highest order; but it does not involve scientific thought, as any astronomer will tell you.

By such calculations the places of the planet Uranus at different times of the year had been predicted and set down. The predictions were not fulfilled. Then arose Adams, and from these errors in the prediction he calculated the place of an entirely new planet, that had never yet been suspected; and you all know how the new planet was actually found in that place. Now this prediction does involve scientific thought, as anyone who has studied it will tell you.

Here then are two cases of thought about the same

subject, both predicting events by the application of previous experience, yet we say one is *technical* and the other *scientific*.

Now let us take an example from the building of bridges and roofs. When an opening is to be spanned over by a material construction, which must bear a certain weight without bending enough to injure itself, there are two forms in which this construction can be made, the arch and the chain. Every part of an arch is compressed or pushed by the other parts; every part of a chain is in a state of tension, or is pulled by the other parts. In many cases these forms are united, A girder consists of two main pieces or booms, of which the upper one acts as an arch and is compressed, while the lower one acts as a chain and is pulled; and this is true even when both the pieces are quite straight. They are enabled to act in this way by being tied together, or braced, as it is called, by cross pieces, which you must often have seen. Now suppose that any good practical engineer makes a bridge or roof upon some approved pattern which has been made before. He designs the size and shape of it to suit the opening which has to be spanned; selects his material according to the locality; assigns the strength which must be given to the several parts of the structure according to the load which it will have to bear. There is a great deal of thought in the making of this design, whose success is predicted by the application of previous experience; it requires technical skill of a very high order; but it is not scientific thought. On the other hand, Mr. Fleeming Jenkin[1] designs a

[1] *On Braced Arches and Suspension Bridges.* Edinburgh: Neill, 1870.

roof consisting of two arches braced together, instead
of an arch and a chain braced together; and although
this form is quite different from any known structure,
yet before it is built he assigns with accuracy the
amount of material that must be put into every part
of the structure in order to make it bear the required
load, and this prediction may be trusted with perfect
security.   What is the natural comment on this?
Why, that Mr. Fleeming Jenkin is a scientific engineer.

Now it seems to me that the difference between
scientific and merely technical thought, not only in
these but in all other instances which I have considered, is just this: Both of them make use of experience to direct human action; but while technical
thought or skill enables a man to deal with the same
circumstances that he has met with before, scientific
thought enables him to deal with different circumstances that he has never met with before.   But how
can experience of one thing enable us to deal with
another quite different thing?   To answer this question we shall have to consider more closely the nature
of scientific thought.

Let us take another example.   You know that if
you make a dot on a piece of paper, and then hold a
piece of Iceland spar over it, you will see not one dot
but two.   A mineralogist, by measuring the angles of
a crystal, can tell you whether or no it possesses this
property without looking through it.   He requires no
scientific thought to do that.   But Sir William Rowan
Hamilton, the late Astronomer Royal of Ireland, knowing these facts and also the explanation of them which

Fresnel had given, thought about the subject, and he predicted that by looking through certain crystals in a particular direction we should see not two dots but a continuous circle. Mr. Lloyd made the experiment, and saw the circle, a result which had never been even suspected. This has always been considered one of the most signal instances of scientific thought in the domain of physics. It is most distinctly an application of experience gained under certain circumstances to entirely different circumstances.

Now suppose that the night before coming down to Brighton you had dreamed of a railway accident caused by the engine getting frightened at a flock of sheep and jumping suddenly back over all the carriages; the result of which was that your head was unfortunately cut off, so that you had to put it in your hat-box and take it back home to be mended. There are, I fear, many persons even at this day, who would tell you that after such a dream it was unwise to travel by railway to Brighton. This is a proposal that you should take experience gained while you are asleep, when you have no common sense,—experience about a phantom railway, and apply it to guide you when you are awake and have common sense, in your dealings with a real railway. And yet this proposal is not dictated by scientific thought.

Now let us take the great example of Biology. I pass over the process of classification, which itself requires a great deal of scientific thought; in particular when a naturalist who has studied and monographed a fauna or a flora rather than a family is able at once to pick out the distinguishing characters required for the subdivision of an order quite new to him. Suppose that we possess all this minute and comprehensive knowledge of plants and animals and intermediate organisms, their affinities and differences, their structures and functions;—a vast body of experience, collected by incalculable labour and devotion. Then comes Mr. Herbert Spencer: he takes that experience of life which is not human, which is apparently stationary, going on in exactly the same way from year to year, and he applies that to tell us how to deal with the changing characters of human nature and human society. How is it that experience of this sort, vast as it is, can guide us in a matter so different from itself? How does scientific thought, applied to the development of a kangaroo fœtus or the movement of the sap in exogens, make prediction possible for the first time in that most important of all sciences, the relations of man with man?

In the dark or unscientific ages men had another way of applying experience to altered circumstances. They believed, for example, that the plant called Jew's-ear, which does bear a certain resemblance to the human ear, was a useful cure for diseases of that organ. This doctrine of ' signatures,' as it was called, exercised an enormous influence on the medicine of the time. I need hardly tell you that it is hopelessly unscientific; yet it agrees with those other examples that we have been considering in this particular; that it applies experience about the shape of a plant—which is one circumstance connected with it—to dealings with its medicinal properties, which are other and different circumstances. Again, suppose that you had been frightened by a thunder-

storm on land, or your heart had failed you in a storm
at sea; if anyone then told you that in consequence of
this you should always cultivate an unpleasant sensation
in the pit of your stomach, till you took delight in it,
that you should regulate your sane and sober life by
the sensations of a moment of unreasoning terror : this
advice would not be an example of scientific thought.
Yet it would be an application of past experience to new
and different circumstances.

But you will already have observed what is the
additional clause that we must add to our definition in
order to describe scientific thought and that only. The
step between experience about animals and dealings
with changing humanity is the law of evolution. The
step from errors in the calculated places of Uranus to
the existence of Neptune is the law of gravitation.
The step from the observed behaviour of crystals to
conical refraction is made up of laws of light and geometry. The step from old bridges to new ones is the
laws of elasticity and the strength of materials.

The step, then, from past experience to new circumstances must be made in accordance with an observed
uniformity in the order of events. This uniformity has
held good in the past in certain places ; if it should also
hold good in the future and in other places, then, being
combined with our experience of the past, it enables us
to predict the future, and to know what is going on
elsewhere ; so that we are able to regulate our conduct
in accordance with this knowledge.

The aim of scientific thought, then, is to apply past
experience to new circumstances ; the instrument is an
observed uniformity in the course of events. By the

use of this instrument it give us information transcending our experience, it enables us to infer things that we have not seen from things that we have seen; and the evidence for the truth of that information depends on our supposing that the uniformity holds good beyond our experience. I now want to consider this uniformity a little more closely; to show how the character of scientific thought and the force of its inferences depend upon the character of the uniformity of Nature. I cannot of course tell you all that is known of this character without writing an encyclopædia; but I shall confine myself to two points of it about which it seems to me that just now there is something to be said. I want to find out what we mean when we say that the uniformity of Nature is *exact*; and what we mean when we say that it is *reasonable*.

When a student is first introduced to those sciences which have come under the dominion of mathematics, a new and wonderful aspect of Nature bursts upon his view. He has been accustomed to regard things as essentially more or less vague. All the facts that he has hitherto known have been expressed qualitatively, with a little allowance for error on either side. Things which are let go fall to the ground. A very observant man may know also that they fall faster as they go along. But our student is shown that, after falling for one second in a vacuum, a body is going at the rate of thirty-two feet per second, that after falling for two seconds it is going twice as fast, after going two and a half seconds two and a half times as fast. If he makes the experiment, and finds a single inch per second too much or too little in the rate, one of two things must have happened : either the law of falling bodies has been wrongly stated, or the experiment is not accurate—there is some mistake. He finds reason to think that the latter is always the case ; the more carefully he goes to work, the more of the error turns out to belong to the experiment. Again, he may know that water consists of two gases, oxygen and hydrogen, combined ; but he now learns that two pints of steam at a temperature of 150° Centigrade will always make two pints of hydrogen and one pint of oxygen at the same temperature, all of them being pressed as much as the atmosphere is pressed. If he makes the experiment and gets rather more or less than a pint of oxygen, is the law disproved? No ; the steam was impure, or there was some mistake. Myriads of analyses attest the law of combining volumes ; the more carefully they are made, the more nearly they coincide with it. The aspects of the faces of a crystal are connected together by a geometrical law, by which, four of them being given, the rest can be found. The place of a planet at a given time is calculated by the law of gravitation ; if it is half a second wrong, the fault is in the instrument, the observer, the clock, or the law ; now, the more observations are made, the more of this fault is brought home to the instrument, the observer, and the clock. It is no wonder, then, that our student, contemplating these and many like instances, should be led to say, 'I have been shortsighted ; but I have now put on the spectacles of science which Nature had prepared for my eyes ; I see that things have definite outlines, that the world is ruled by exact and rigid mathematical laws ; καὶ σύ, θεός, γεωμετρεῖς.' It is our business to consider whether he is right in so concluding. Is the uniformity of Nature absolutely exact, or only more exact than our experiments?

At this point we have to make a very important distinction. There are two ways in which a law may be inaccurate. The first way is exemplified by that law of Galileo which I mentioned just now: that a body falling *in vacuo* acquires equal increase in velocity in equal times. No matter how many feet per second it is going, after an interval of a second it will be going thirty-two *more* feet per second. We now know that this rate of increase is not exactly the same at different heights, that it depends upon the distance of the body from the centre of the earth; so that the law is only approximate; instead of the increase of velocity being exactly *equal* in equal times, it itself increases very slowly as the body falls. We know also that this variation of the law from the truth is *too small to be perceived* by direct observation on the change of velocity. But suppose we have invented means for observing this, and have verified that the increase of velocity is inversely as the squared distance from the earth's centre. Still the law is not accurate; for the earth does not attract accurately towards her centre, and the direction of attraction is continually varying with the motion of the sea; the body will not even fall in a straight line. The sun and the planets, too, especially the moon, will produce deviations; yet the sum of all these errors will escape our new process of observation, by being a great deal smaller than the necessary errors of that observation. But when these again have been allowed for, there is still the influence of the stars. In this case, however, we only give up one exact law for another. It may still be held that if the effect of every particle of matter in the universe on the falling body were calculated according to the law of gravitation, the body would move exactly as this calculation required. And if it were objected that the body must be slightly magnetic or diamagnetic, while there are magnets not an infinite way off; that a very minute repulsion, even at sensible distances, accompanies the attraction; it might be replied that these phenomena are themselves subject to exact laws, and that when *all* the laws have been taken into account, the actual motion will exactly correspond with the calculated motion.

I suppose there is hardly a physical student (unless he has specially considered the matter) who would not at once assent to the statement I have just made; that if we knew all about it, Nature would be found universally subject to exact numerical laws. But let us just consider for another moment what this means.

The word 'exact' has a practical and a theoretical meaning. When a grocer weighs you out a certain quantity of sugar very carefully, and says it is exactly a pound, he means that the difference between the mass of the sugar and that of the pound weight he employs is too small to be detected by his scales. If a chemist had made a special investigation, wishing to be as accurate as he could, and told you this was exactly a pound of sugar, he would mean that the mass of the sugar differed from that of a certain standard piece of platinum by a quantity too small to be detected by *his* means of weighing, which are a thousandfold more accurate than the grocer's. But what would a mathematician mean, if he made the same statement? He would mean this. Suppose the mass of the standard pound to be represented by a length, say a foot, measured on a certain line; so that half a pound would be represented by six inches, and so on. And let the difference between the mass of the sugar and that of the standard pound be drawn upon the same line to the same scale. Then, if that difference were magnified an infinite number of times, it would still be invisible. This is the theoretical meaning of exactness; the practical meaning is only very close approximation; *how* close, depends upon the circumstances. The knowledge then of an exact law in the theoretical sense would be equivalent to an infinite observation. I do not say that such knowledge is impossible to man; but I do say that it would be absolutely different in kind from any knowledge that we possess at present.

I shall be told, no doubt, that we do possess a great deal of knowledge of this kind, in the form of geometry and mechanics; and that it is just the example of these sciences that has led men to look for exactness in other quarters. If this had been said to me in the last century, I should not have known what to reply. But it happens that about the beginning of the present century the foundations of geometry were criticised independently by two mathematicians, Lobatschewsky[1] and the immortal Gauss;[2] whose results have been extended and generalized more recently by Riemann[3]

---

[1] *Geometrische Untersuchungen zur Theorie der Parallellinien.* Berlin, 1840. Translated by Hoüel. Gauthier-Villars, 1866.

[2] Letter to Schumacher, Nov. 28, 1846 (refers to 1792).

[3] *Ueber die Hypothesen welche der Geometrie zu Grunde liegen.* Göttingen, Abhandl., 1866–7. Translated by Hoüel in *Annali di Matematica*, Milan, vol. iii.

and Helmholtz.[1] And the conclusion to which these investigations lead is that, although the assumptions which were very properly made by the ancient geometers are practically exact—that is to say, more exact than experiment can be—for such finite things as we have to deal with, and such portions of space as we can reach ; yet the truth of them for very much larger things, or very much smaller things, or parts of space which are at present beyond our reach, is a matter to be decided by experiment, when its powers are considerably increased. I want to make as clear as possible the real state of this question at present, because it is often supposed to be a question of words or metaphysics, whereas it is a very distinct and simple question of fact. I am supposed to know then that the three angles of a rectilinear triangle are exactly equal to two right angles. Now suppose that three points are taken in space, distant from one another as far as the Sun is from *a* Centauri, and that the shortest distances between these points are drawn so as to form a triangle. And suppose the angles of this triangle to be very accurately measured and added together ; this can at present be done so accurately that the error shall certainly be less than one minute, less therefore than the five-thousandth part of a right angle. Then I do not know that this sum would differ at all from two right angles ; but also I do not know that the difference would be less than ten degrees, or the ninth part of a right angle.[2] And I have reasons for not knowing.

---

[1] *The Axioms of Geometry*, Academy, vol. i. p. 128 (a popular exposition). [And see now his article in *Mind*, No. III.].

[2] Assuming that parallax observations prove the deviation less than half a second for a triangle whose vertex is at the star and base a diameter of the earth's orbit.

This example is exceedingly important as showing the connexion between exactness and universality. It is found that the deviation if it exists must be nearly proportional to the area of the triangle. So that the error in the case of a triangle whose sides are a mile long would be obtained by dividing that in the case I have just been considering by four hundred quadrillions; the result must be a quantity inconceivably small, which no experiment could detect. But between this inconceivably small error and no error at all, there is fixed an enormous gulf; the gulf between practical and theoretical exactness, and, what is even more important, the gulf between what is practically universal and what is theoretically universal. I say that a law is practically universal which is more exact than experiment for all cases that might be got at by such experiments as we can make. We assume this kind of universality, and we find that it pays us to assume it. But a law would be theoretically universal if it were true of all cases whatever; and this is what we do not know of any law at all.

I said there were two ways in which a law might be inexact. There is a law of gases which asserts that when you compress a perfect gas the pressure of the gas increases exactly in the proportion in which the volume diminishes. Exactly; that is to say, the law is more accurate than the experiment, and experiments are corrected by means of the law. But it so happens that this law has been explained; we know precisely what it is that happens when a gas is compressed. We know that a gas consists of a vast number of separate molecules, rushing about in all directions with all manner of velocities, but so that the mean velocity of the molecules of air in this room, for example, is about twenty miles a minute. The pressure of the gas on any surface with which it is in contact is nothing more than the impact of these small particles upon it. On any surface large enough to be seen there are millions of these impacts in a second. If the space in which the gas is confined be diminished, the average rate at which the impacts take place will be increased in the same proportion; and because of the enormous number of them, the actual rate is always exceedingly close to the average. But the law is one of statistics; its accuracy depends on the enormous numbers involved; and so, from the nature of the case, its exactness cannot be theoretical or absolute.

Nearly all the laws of gases have received these statistical explanations; electric and magnetic attraction and repulsion have been treated in a similar manner; and an hypothesis of this sort has been suggested even for the law of gravity. On the other hand the manner in which the molecules of a gas interfere with each other proves that they repel one another inversely as the fifth power of the distance;[1] so that we here find at the basis of a statistical explanation a law which has the form of theoretical exactness. Which of these forms is to win? It seems to me again that we do not know, and that the recognition of our ignorance is the surest way to get rid of it.

The world in general has made just the remark that I have attributed to a fresh student of the applied

---

[1] [This statement of the law has since been abandoned: see ‘The Unseen Universe,’ below.]

sciences. As the discoveries of Galileo, Kepler, Newton, Dalton, Cavendish, Gauss, displayed ever new
phenomena following mathematical laws, the theoretical exactness of the physical universe was taken for
granted. Now, when people are hopelessly ignorant of
a thing, they quarrel about the source of their knowledge. Accordingly many maintained that we know
these exact laws by intuition. These said always one
true thing, that we did not know them from experience.
Others said that they were really given in the facts, and
adopted ingenious ways of hiding the gulf between the
two. Others again deduced from transcendental considerations sometimes the laws themselves, and sometimes what through imperfect information they supposed
to be the laws. But more serious consequences arose
when these conceptions derived from Physics were
carried over into the field of Biology. Sharp lines of
division were made between kingdoms and classes and
orders; an animal was described as a miracle to the
vegetable world; specific differences which are practically
permanent within the range of history were regarded
as permanent through all time ; a sharp line was drawn
between organic and inorganic matter. Further investigation, however, has shown that accuracy had been
prematurely attributed to the science, and has filled up
all the gulfs and gaps that hasty observers had invented.
The animal and vegetable kingdoms have a debateable
ground between them, occupied by beings that have
the characters of both and yet belong distinctly to
neither. Classes and orders shade into one another all
along their common boundary. Specific differences
turn out to be the work of time. The line dividing

organic matter from inorganic, if drawn to-day, must be moved to-morrow to another place ; and the chemist will tell you that the distinction has now no place in his science except in a technical sense for the convenience of studying carbon compounds by themselves. In Geology the same tendency gave birth to the doctrine of distinct periods, marked out by the character of the strata deposited in them all over the sea ; a doctrine than which, perhaps, no ancient cosmogony has been further from the truth, or done more harm to the progress of science. Refuted many years ago by Mr. Herbert Spencer,[1] it has now fairly yielded to an attack from all sides at once, and may be left in peace.

When then we say that the uniformity which we observe in the course of events is exact and universal, we mean no more than this: that we are able to state general rules which are far more exact than direct experiment, and which apply to all cases that we are at present likely to come across. It is important to notice, however, the effect of such exactness as we observe upon the nature of inference. When a telegram arrived stating that Dr. Livingstone had been found by Mr. Stanley, what was the process by which you inferred the finding of Dr. Livingstone from the appearance of the telegram? You assumed over and over again the existence of uniformity in nature. That the newspapers had behaved as they generally do in regard to telegraphic messages ; that the clerks had followed the known laws of the action of clerks ; that electricity had behaved in the cable exactly as it behaves in the

---

[1] 'Illogical Geology,' in *Essays*, vol. i. Originally published in 1859,

laboratory; that the actions of Mr. Stanley were related to his motives by the same uniformities that affect the actions of other men; that Dr. Livingstone's handwriting conformed to the curious rule by which an ordinary man's handwriting may be recognized as having persistent characteristics even at different periods of his life. But you had a right to be much more sure about some of these inferences than about others. The law of electricity was known with practical exactness, and the conclusions derived from it were the surest things of all. The law about the handwriting, belonging to a portion of physiology which is unconnected with consciousness, was known with less, but still with considerable accuracy. But the laws of human action in which consciousness is concerned are still so far from being completely analysed and reduced to an exact form that the inferences which you made by their help were felt to have only a provisional force. It is possible that by-and-by, when psychology has made enormous advances and become an exact science, we may be able to give to testimony the sort of weight which we give to the inferences of physical science. It will then be possible to conceive a case which will show how completely the whole process of inference depends on our assumption of uniformity. Suppose that testimony, having reached the ideal force I have imagined, were to assert that a certain river runs uphill. You could infer nothing at all. The arm of inference would be paralysed, and the sword of truth broken in its grasp; and reason could only sit down and wait until recovery restored her limb, and further experience gave her new weapons.

I want in the next place to consider what we mean when we say that the uniformity which we have observed in the course of events is *reasonable* as well as exact.

No doubt the first form of this idea was suggested by the marvellous adaptation of certain natural structures to special functions. The first impression of those who studied comparative anatomy was that every part of the animal frame was fitted with extraordinary completeness for the work that it had to do. I say extraordinary, because at the time the most familiar examples of this adaptation were manufactures produced by human ingenuity; and the completeness and minuteness of natural adaptations were seen to be far in advance of these. The mechanism of limbs and joints was seen to be adapted, far better than any existing ironwork, to those motions and combinations of motion which were most useful to the particular organisms. The beautiful and complicated apparatus of sensation caught up indications from the surrounding medium, sorted them, analysed them, and transmitted the results to the brain in a manner with which, at the time I am speaking of, no artificial contrivance could compete. Hence the belief grew amongst physiologists that every structure which they found must have its function and subserve some useful purpose; a belief which was not without its foundation in fact, and which certainly (as Dr. Whewell remarks) has done admirable service in promoting the growth of physiology. Like all beliefs found successful in one subject, it was carried over into another, of which a notable example is given in the speculations of Count Rumford about the physical properties of water. Pure water attains its greatest density at a temperature of about $39\frac{1}{2}°$ Fahrenheit; it expands and becomes lighter whether it is cooled or heated, so as to alter that temperature. Hence it was concluded that water in this state must be at the bottom of the sea, and that by such means the sea was kept from freezing all through; as it was supposed must happen if the greatest density had been that of ice. Here then was a substance whose properties were eminently adapted to secure an end essential to the maintenance of life upon the earth. In short, men came to the conclusion that the order of nature was reasonable in the sense that everything was adapted to some good end.

Further consideration, however, has led men out of that conclusion in two different ways. First, it was seen that the facts of the case had been wrongly stated. Cases were found of wonderfully complicated structures that served no purpose at all; like the teeth of that whale of which you heard in Section D the other day, or of the Dugong, which has a horny palate covering them all up and used instead of them; like the eyes of the unborn mole, that are never used, though perfect as those of a mouse until the skull opening closes up, cutting them off from the brain, when they dry up and become incapable of use; like the outsides of your own ears, which are absolutely of no use to you. And when human contrivances were more advanced it became clear that the natural adaptations were subject to criticism. The eye, regarded as an optical instrument of human manufacture, was thus described by Helmholtz—the physiologist who learned physics for the sake of his physiology, and mathematics for the sake of his physics, and is now in the first rank of all three. He said, 'If an optician sent me that as an instrument, I should send it back to him with grave reproaches for the carelessness of his work, and demand the return of my money.'

The extensions of the doctrine into Physics were found to be still more at fault. That remarkable property of pure water, which was to have kept the sea from freezing, does not belong to salt water, of which the sea itself is composed. It was found, in fact, that the idea of a reasonable adaptation of means to ends, useful as it had been in its proper sphere, could yet not be called universal, or applied to the order of nature as a whole.

Secondly, this idea has given way because it has been superseded by a higher and more general idea of what is reasonable, which has the advantage of being applicable to a large portion of physical phenomena besides. Both the adaptation and the non-adaptation which occur in organic structures have been *explained*. The scientific thought of Dr. Darwin, of Mr. Herbert Spencer, and of Mr. Wallace, has described that hitherto unknown process of adaptation as consisting of perfectly well-known and familiar processes. There are two kinds of these: the direct processes, in which the physical changes required to produce a structure are worked out by the very actions for which that structure becomes adapted—as the backbone or notochord has been modified from generation to generation by the bendings which it has undergone; and the indirect processes included under the head of Natural Selection—the reproduction of children slightly different from their parents, and the survival of those which are best fitted to hold their own in the struggle for existence. Naturalists might give you some idea of the rate at which we are getting explanations of the evolution of all parts of animals and plants—the growth of the skeleton, of the nervous system and its mind, of leaf and flower. But what then do we mean by *explanation*?

We were considering just now an explanation of a law of gases—the law according to which pressure increases in the same proportion in which volume diminishes. The explanation consisted in supposing that a gas is made up of a vast number of minute particles always flying about and striking against one another, and then showing that the rate of impact of such a crowd of particles on the sides of the vessel containing them would vary exactly as the pressure is found to vary. Suppose the vessel to have parallel sides, and that there is only one particle rushing backwards and forwards between them; then it is clear that if we bring the sides together to half the distance, the particle will hit each of them twice as often, or the pressure will be doubled. Now it turns out that this would be just as true for millions of particles as for one, and when they are flying in all directions instead of only in one direction and its opposite. Observe now; it is a perfectly well-known and familiar thing that a body should strike against an opposing surface and bound off again; and it is a mere everyday occurrence that what has only half so far to go should be back in half the time; but that pressure should be strictly proportional to density is a comparatively strange, unfamiliar phenomenon. The explanation describes the unknown and unfamiliar as being made up of the known and the familiar ; and this, it seems to me, is the true meaning of explanation.[1]

Here is another instance. If small pieces of camphor are dropped into water, they will begin to spin round and swim about in a most marvellous way. Mr. Tomlinson gave, I believe, the explanation of this. We must observe, to begin with, that every liquid has a skin which holds it ; you can see that to be true in the case of a drop, which looks as if it were held in a bag. But the tension of this skin is greater in some liquids than in others ; and it is greater in camphor and water than in pure water. When the camphor is dropped into water it begins to dissolve and get surrounded with camphor and water instead of water. If the fragment of camphor were exactly symmetrical, nothing more would happen ; the tension would be greater in its immediate neighbourhood, but no motion would follow. The camphor, however, is irregular in shape ; it dissolves more on one side than the other ; and consequently gets pulled about, because the tension of the skin is greater where the camphor is most dissolved. Now it is probable that this is not nearly so satisfactory an explanation to you as it was to me when I was first told of it ; and for this reason. By that time I was already perfectly familiar with the notion of a skin upon the surface of liquids, and I had been taught by

---

[1] This view differs from those of Mr. J. S. Mill and Mr. Herbert Spencer in requiring every explanation to contain an addition to our knowledge about the thing explained. Both those writers regard subsumption under a general law as a species of explanation. See also Ferrier's 'Remains,' vol. ii. p. 436.

means of it to work out problems in capillarity. The explanation was therefore a description of the unknown phenomenon which I did not know how to deal with as made up of known phenomena which I did know how to deal with. But to many of you possibly the liquid skin may seem quite as strange and unaccountable as the motion of camphor on water.

And this brings me to consider the source of the pleasure we derive from an explanation. By known and familiar I mean that which we know how to deal with, either by action in the ordinary sense, or by active thought. When therefore that which we do not know how to deal with is described as made up of things that we do know how to deal with, we have that sense of increased power which is the basis of all higher pleasures. Of course we may afterwards by association come to take pleasure in explanation for its own sake. Are we then to say that the observed order of events is reasonable, in the sense that all of it admits of explanation? That a process may be capable of explanation, it must break up into simpler constituents which are already familiar to us. Now, first, the process may itself be simple, and not break up; secondly, it may break up into elements which are as unfamiliar and impracticable as the original process.

It is an explanation of the moon's motion to say that she is a falling body, only she is going so fast and is so far off that she falls quite round to the other side of the earth, instead of hitting it ; and so goes on for ever. But it is no explanation to say that a body falls because of gravitation. That means that the motion of the body may be resolved into a motion of every one of its particles towards every one of the particles of the earth, with an acceleration inversely as the square of the distance between them. But this attraction of two particles must always, I think, be less familiar than the original falling body, however early the children of the future begin to read their Newton. Can the attraction itself be explained? Le Sage said that there is an everlasting hail of innumerable small ether-particles from all sides, and that the two material particles shield each other from this and so get pushed together. This is an explanation; it may or may not be a true one. The attraction may be an ultimate simple fact; or it may be made up of simpler facts utterly unlike anything that we know at present; and in either of these cases there is no explanation. We have no right to conclude, then, that the order of events is always capable of being explained.

There is yet another way in which it is said that Nature is reasonable; namely, inasmuch as every effect has a cause. What do we mean by this?

In asking this question, we have entered upon an appalling task. The word represented by 'cause' has sixty-four meanings in Plato and forty-eight in Aristotle. These were men who liked to know as near as might be what they meant; but how many meanings it has had in the writings of the myriads of people who have not tried to know what they meant by it will, I hope, never be counted. It would not only be the height of presumption in me to attempt to fix the meaning of a word which has been used by so grave authority in so many and various senses; but it would seem a thankless task to do that once more which has been done so often at

sundry times and in divers manners before.   And yet
without this we cannot determine what we mean by
saying that the order of nature is reasonable.   I shall
evade the difficulty by telling you Mr. Grote's opinion.[1]
You come to a scarecrow and ask, what is the cause of
this?   You find that a man made it to frighten the
birds.   You go away and say to yourself, 'Everything
resembles this scarecrow.   Everything has a purpose.'
And from that day the word 'cause' means for you
what Aristotle meant by 'final cause.'   Or you go into
a hairdresser's shop, and wonder what turns the wheel
to which the rotatory brush is attached.   On investigating other parts of the premises, you find a man working away at a handle.   Then you go away and say,
'Everything is like that wheel.   If I investigated
enough, I should always find a man at a handle.'   And
the man at the handle, or whatever corresponds to him,
is from henceforth known to you as 'cause.'

And so generally.   When you have made out any
sequence of events to your entire satisfaction, so that
you know all about it, the laws involved being so
familiar that you seem to see how the beginning must
have been followed by the end, then you apply that as
a simile to all other events whatever, and your idea of
cause is determined by it.   Only when a case arises, as
it always must, to which the simile will not apply, you
do not confess to yourself that it was only a simile and
need not apply to everything, but you say, 'The cause
of that event is a mystery which must remain for ever
unknown to me.'   On equally just grounds the nervous
system of my umbrella is a mystery which must remain

[1] Plato, vol. ii. (Phædo).

for ever unknown to me. My umbrella has no nervous system; and the event to which your simile did not apply has no cause in your sense of the word. When we say then that every effect has a cause, we mean that every event is connected with something in a way that might make somebody call that the cause of it. But I, at least, have never yet seen any single meaning of the word that could be fairly applied to the *whole* order of nature.

From this remark I cannot even except an attempt recently made by Mr. Bain to give the word a universal meaning, though I desire to speak of that attempt with the greatest respect. Mr. Bain[1] wishes to make the word ' cause ' hang on in some way to what we call the law of energy; but though I speak with great diffidence I do think a careful consideration will show that the introduction of this word ' cause' can only bring confusion into a matter which is distinct and clear enough to those who have taken the trouble to understand what energy means. It would be impossible to explain that this evening; but I may mention that ' energy' is a technical term out of mathematical physics, which requires of most men a good deal of careful study to understand it accurately.

Let us pass on to consider, with all the reverence which it demands, another opinion held by great numbers of the philosophers who have lived in the Brightening Ages of Europe; the opinion that at the basis of the natural order there is something which we can know to be *unreasonable*, to evade the processes of human thought. The opinion is set forth first by Kant,

[1] *Inductive Logic*, chap. iv.

so far as I know, in the form of his famous doctrine of the antinomies or contradictions, a later form [1] of which I will endeavour to explain to you. It is said, then, that space must either be infinite or have a boundary. Now you cannot conceive infinite space; and you cannot conceive that there should be any end to it. Here then, are two things, one of which must be true, while each of them is inconceivable; so that our thoughts about space are hedged in, as it were, by a contradiction. Again, it is said that matter must either be infinitely divisible, or must consist of small particles incapable of further division. Now you cannot conceive a piece of matter divided into an infinite number of parts, while, on the other hand, you cannot conceive a piece of matter, however small, which absolutely cannot be divided into two pieces; for, however great the forces are which join the parts of it together, you can imagine stronger forces able to tear it in pieces. Here, again, there are two statements, one of which must be true, while each of them is separately inconceivable; so that our thoughts about matter also are hedged in by a contradiction. There are several other cases of the same thing, but I have selected these two as instructive examples. And the conclusion to which philosophers were led by the contemplation of them was that on every side, when we approach the limits of existence, a contradiction must stare us in the face. The doctrine has been developed and extended by the great successors of Kant; and this unreasonable, or

---

[1] That of Mr. Herbert Spencer, *First Principles*. I believe Kant himself would have admitted that the antinomies do not exist for the empiricist. [Much less does he say that either of a pair of antinomies must be true. The real Kantian position is that both assertions are illegitimate.]

unknowable, which is also called the absolute and the unconditioned, has been set forth in various ways as that which we know to be the true basis of all things. As I said before, I approach this doctrine with all the reverence which should be felt for that which has guided the thoughts of so many of the wisest of mankind. Nevertheless I shall endeavour to show that in these cases of supposed contradiction there is always something which we do not know now, but of which we cannot be sure that we shall be ignorant next year. The doctrine is an attempt to found a positive statement upon this ignorance, which can hardly be regarded as justifiable. Spinoza said, ' A free man thinks of nothing so little as of death ; ' it seems to me we may parallel this maxim in the case of thought, and say, ' A wise man only remembers his ignorance in order to destroy it.' A boundary is that which divides two adjacent portions of space. The question, then, ' Has space (in general) a boundary ? ' involves a contradiction in terms, and is, therefore, unmeaning. But the question, ' Does space contain a finite number of cubic miles, or an infinite number ? ' is a perfectly intelligible and reasonable question which remains to be answered by experiment.[1] The surface of the sea would still contain a finite number of square miles, if there were no land to bound it. Whether or no the space in which we live is of this nature remains to be seen. If its extent is finite, we may quite possibly be able to assign that extent next year ; if, on the other hand, it has no end, it is true that the knowledge of that fact would be quite different

---

[1] The very important distinction between *unboundedness* and *infinite extent* is made by Riemann, loc. cit.

from any knowledge we at present possess, but we have no right to say that such knowledge is impossible. Either the question will be settled once for all, or the extent of space will be shown to be greater than a quantity which will increase from year to year with the improvement of our sources of knowledge. Either alternative is perfectly conceivable, and there is no contradiction. Observe especially that the supposed contradiction arises from the assumption of theoretical exactness in the laws of geometry. The other case that I mentioned has a very similar origin. The idea of a piece of matter the parts of which are held together by forces, and are capable of being torn asunder by greater forces, is entirely derived from the large pieces of matter which we have to deal with. We do not know whether this idea applies in any sense even to the *molecules* of gases; still less can we apply it to the *atoms* of which they are composed. The word force is used of two phenomena: the pressure, which when two bodies are in contact connects the motion of each with the position of the other; and attraction or repulsion,—that is to say, a change of velocity in one body ·depending on the position of some other body which is not in contact with it. We do not know that there is anything corresponding to either of these phenomena in the case of a molecule. A meaning can, however, be given to the question of the divisibility of matter in this way. We may ask if there is any piece of matter so small that its properties as matter depend upon its remaining all in one piece. This question is reasonable; but we cannot answer it at present, though we are not at all sure that we shall be equally ignorant next year. If there is no such piece of matter, no such limit to the division which shall leave it matter, the knowledge of that fact would be different from any of our present knowledge; but we have no right to say that it is impossible. If, on the other hand, there *is* a limit, it is quite possible that we may have measured it by the time the Association meets at Bradford. Again, when we are told that the infinite extent of space, for example, is something that we cannot conceive at present, we may reply that this is only natural, since our experience has never yet supplied us with the means of conceiving such things. But then we cannot be sure that the facts will not make us learn to conceive them; in which case they will cease to be inconceivable. In fact, the putting of limits to human conception must always involve the assumption that our previous experience is universally valid in a theoretical sense; an assumption which we have already seen reason to reject. Now you will see that our consideration of this opinion has led us to the true sense of the assertion that the Order of Nature is reasonable. If you will allow me to define a reasonable question as one which is asked in terms of ideas justified by previous experience, without itself contradicting that experience, then we may say, as the result of our investigation, that to every reasonable question there is an intelligible answer which either we or posterity may know.

We have, then, come somehow to the following conclusions. By scientific thought we mean the application of past experience to new circumstances by means of an observed order of events. By saying that this order of events is exact we mean that it is exact enough to correct experiments by, but we do not mean that it is

theoretically or absolutely exact, because we do not
know.   The process of inference we found to be in itself
an assumption of uniformity, and we found that, as the
known exactness of the uniformity became greater, the
stringency of the inference increased.   By saying that
the order of events is reasonable we do not mean that
everything has a purpose, or that everything can be
explained, or that everything has a cause; for neither
of these is true.   But we mean that to every reasonable
question there is an intelligible answer, which either we
or posterity may know *by the exercise of scientific thought.*

For I specially wish you not to go away with the
idea that the exercise of scientific thought is properly
confined to the subjects from which my illustrations
have been chiefly drawn to-night.   When the Roman
jurists applied their experience of Roman citizens to
dealings between citizens and aliens, showing by the
difference of their actions that they regarded the circumstances as essentially different, they laid the foundations
of that great structure which has guided the social
progress of Europe.   That procedure was an instance of
strictly scientific thought.   When a poet finds that he
has to move a strange new world which his predecessors
have not moved; when, nevertheless, he catches fire
from their flashes, arms from their armoury, sustenta-
tion from their foot-prints, the procedure by which he
applies old experience to new circumstances is nothing
greater or less than scientific thought.   When the
moralist, studying the conditions of society and the ideas
of right and wrong which have come down to us from
a time when war was the normal condition of man and
success in war the only chance of survival, evolves from

them the conditions and ideas which must accompany a time of peace, when the comradeship of equals is the condition of national success; the process by which he does this is scientific thought and nothing else. Remember, then, that it is the guide of action; that the truth which it arrives at is not that which we can ideally contemplate without error, but that which we may act upon without fear; and you cannot fail to see that scientific thought is not an accompaniment or condition of human progress, but human progress itself. And for this reason the question what its characters are, of which I have so inadequately endeavoured to give you some glimpse, is the question of all questions for the human race.

ATOMS.[1]

IF I were to wet my finger and then rub it along the edge of this glass, I should no doubt persuade the glass to give out a certain musical note. So also if I were to sing to that glass the same note loud enough, I should get the glass to answer me back with a note.

I want you to remember that fact, because it is of capital importance for the arguments we shall have to consider to-night. The very same note which I can get the tumbler to give out by agitating it, by rubbing the edge, that same note I can also get the tumbler to answer back to me when I sing to it. Now, remembering that, please to conceive a rather complicated thing that I am now going to try to describe to you. The same property that belongs to the glass belongs also to a bell which is made out of metal. If that bell is agitated by being struck, or in any other way, it will give out the same sound that it will answer back if you sing that sound to it; but if you sing a different sound to it then it will not answer.

Now suppose that I have several of these metal bells which answer to quite different notes, and that they are all fastened to a set of elastic stalks which spring out of a certain centre to which they are fastened. All these bells, then, are not only fastened to these stalks, but they are held there in such a way that they can spin round upon the points to which they are fastened.

And then the centre to which these elastic stalks are fastened or suspended, you may imagine as able to move in all manner of directions, and that the whole structure made up of these bells and stalks and centre is able to spin round any axis whatever. We must also suppose that there is surrounding this structure a certain framework. We will suppose the framework to be made of some elastic material, so that it is able to be pressed in to a certain extent. Suppose that framework is made of whalebone, if you like. This structure I am going for the present to call an 'atom.' I do not mean to say that atoms are made of a structure like that. I do not mean to say that there is anything in an atom which is in the shape of a bell; and I do not mean to say that there is anything analogous to an elastic stalk in it. But what I mean is this—that an atom is something that is capable of vibrating at certain definite rates ; also that it is capable of other motions of its parts besides those vibrations at certain definite rates ; and also that it is capable of spinning round about any axis. Now by the framework which I suppose to be put round that structure made out of bells and elastic stalks, I mean this —that supposing you had two such structures, then you cannot put them closer together than a certain distance, but they will begin to resist being put close together after you have put them as near as that, and they will push each other away if you attempt to put them closer. That is all I mean then. You must only suppose that that structure is described, and that set of ideas is put together, just for the sake of giving us some definite notion of a thing which has similar properties to that structure. But you must not suppose that there is any special part of an atom which has got a bell-like form, or any part like an elastic stalk made out of whalebone.

Now having got the idea of such a complicated structure, which is capable, as we said, of vibratory motion, and of other sorts of motion, I am going on to explain what is the belief of those people who have studied the subject about the composition of the air which fills this room. The air which fills this room is what is called a gas; but it is not a simple gas; it is a mixture of two different gases, oxygen and nitrogen. What is believed about this air is that it consists of quite distinct portions or little masses of air—that is, of little masses each of which is either oxygen or nitrogen; and that these little masses are perpetually flying about in all directions. The number of them in this room is so great that it strains the powers of our numerical system to count them. They are flying about in all directions and mostly in straight lines, except where they get quite near to one another, and then they rebound and fly off in other directions. Part of these little masses which compose the air are of one sort—they are called oxygen. All those little masses which are called oxygen are alike; they are of the same weight; they have the same rates of vibration; and they go about on the average at a certain rate. The other part of these little masses is called nitrogen, and they have a different weight; but the weight of all the nitrogen masses is the same, as nearly as we can make out. They have again the same rates of vibration; but the rates of vibration that belong to them are different from the rates of vibration that belong to the oxygen masses; and the nitrogen masses go about on the average at a certain rate, but this rate is different from the average rate at which the oxygen masses go about. So then, taking up that structure which I endeavoured to de scribe to you at first, we should represent the state of the air in this room as being made up of such a lot of compound atoms of those structures of bells and stalks, with frameworks round them, that I described to you, being thrown about in all directions with great rapidity, and continually impinging against one another, each flying off in a different direction, so that they would go mostly in straight lines (you must suppose them for a moment not to fall down towards the earth), excepting where they come near enough for their two frameworks to be in contact, and then their frameworks throw them off in different directions: that is a conception of the state of things which actually takes place inside of gas.

Now, the conception which scientific men have of the state of things which takes place inside of a liquid is different from that. We should conceive it in this way: We should suppose that a number of these structures are put so close together that their frameworks are always in contact; and yet they are moving about and rolling among one another, so that no one of them keeps the same place for two instants together, and any one of them is travelling all over the whole space. Inside of this glass, where there is a liquid, all the small particles or molecules are running about among one another, and yet none of them goes for any appreciable

portion of its path in a straight line, because there is
no distance however small that it goes without being in
contact with others all around it ; and the effect of this
contact of the others all around it is that they press
against it and force it out of a straight path. So that
the path of a particle in a liquid is a sort of wavy
path ; it goes in and out in all directions, and a particle
at one part of the liquid will, at a certain time, have
traversed all the different parts one after another.

The conception of what happens inside of a solid
body, say a crystal of salt, is different again from this.
It is supposed that the very small particles which constitute that crystal of salt do not travel about from one
part of the crystal to another, but that each one of them
remains pretty much in the same place.   I say 'pretty
much,' but not exactly, and the motion of it is like this :
Suppose one of my structures, with its framework round
it, to be fastened up by elastic strings, so that one string
goes to the ceiling, and another to the floor, and another
to each wall, so that it is fastened by all these strings.
Then if these strings are stretched, and a particle is
displaced in any way, it will just oscillate about its mean
position, and will not go far away from it ; and if forced
away from that position it will come back again.   That
is the sort of motion that belongs to a particle in the
inside of a solid body.   A solid body, such as a crystal
of salt, is made up, just as a liquid or a gas is made up,
of innumerable small particles, but they are so attached
to one another that each of them can only oscillate about
its mean position.   It is very probable that it is also
able to spin about any axis in that position or near it ;
but it is not able to leave that position finally, and to

go and take up another position in the crystal ; it must stop in or near about the same position.

These, then, are the views which are held by scientific men at present about what actually goes on inside of a gaseous body, or a liquid body, or a solid body   In each case the body is supposed to be made up of a very large number of very small particles ; but in one case these particles are very seldom in contact with one another, that is, very seldom within range of each other's action ; in this case they are during the greater part of the time moving separately along straight lines.   In the case of a liquid they are constantly within the range of each other's action ; but they do not move along straight lines for any appreciable part of the time ; they are always changing their position relatively to the other particles, and one of them gets about from one part of the liquid to another.   In the case of a solid they are always also within the range of each other's action, and they are so much within that range that they are not able to change their relative positions ; and each one of them is obliged to remain in very nearly the same position.

Now what I want to do this evening is to explain to you, so far as I can, the reasons which have led scientific men to adopt these views ; and what I wish especially to impress upon you is this, that what is called the ' atomic theory '—that is, what I have just been explaining—is no longer in the position of a theory, but that such of the facts as I have just explained to you are really things which are definitely known and which are no longer suppositions ; that the arguments by which scientific men have been led to adopt these views are such as, to anybody who fairly considers them, justify that person in believing that the statements are true.

Now first of all I want to explain what the reasons are why we believe that the air consists of separate portions, and that these portions are repetitions of the same structures. That is to say that in the air we have two structures really, each of them a great number of times repeated. Take a simple illustration, which is a rather easier one to consider. Suppose we take a vessel which is filled with oxygen. I want to show what the reasons are which lead us to believe that that gas consists of a certain structure which is a great number of times repeated, and that between two examples of that structure which exist inside of the vessel there is a certain empty space which does not contain any oxygen. That oxygen gas contained in the vessel is made up of small particles which are not close together, and each of these particles has a certain structure, which structure also belongs to the rest of the particles. This argument is rather a difficult one, and I shall ask you therefore to follow it as closely as possible, because it is an extremely complicated argument to follow out the first time that it is presented to you.

I want to consider again the case of this finger-glass. You must often have tried that experiment—that a glass will give out when it is agitated the same note which it will return when it is sung to. Well, now, suppose that I have got this room filled with a certain number of such atomic structures as I have endeavoured to describe— that is to say, of sets of bells, the bells answering to certain given notes. Each of these little structures is exactly alike, that is to say, it contains just the bells

corresponding to the same notes. Well, now, suppose
that you sing to a glass or to a bell, there are three
things that may happen. First, you may sing a note
which does not belong to the bell at all. In that case
the bell will not answer; it will not be affected or
agitated by your singing that note, but it will remain
quite still. Next, if you sing a note that belongs to the
bell, but if you sing it rather low, then the effect of that
note will be to make the bell move a little, but the bell
will not move so much as to give back the note in an
audible form. Thirdly, if you sing the note which
belongs to the bell loud enough, then you will so far
agitate the bell that it will give back the note to you
again. Now exactly that same property belongs to a
stretched string or the string of a piano. You know
that if you sing a certain note in a room where there is
a piano, the string belonging to that note will answer
you if you sing loud enough. The other strings won't
answer at all. If you don't sing loud enough the string
will be affected, but not enough to answer you. Now
let us imagine a screen of piano strings, all of exactly
the same length, of the same material, and stretched
equally, and that this screen of strings is put across the
room; that I am at one end and that you are at another,
and that I proceed to sing notes straight up the scale.
While I sing notes which are different from that note
which belongs to the screen of strings, they will pass
through the screen without being altered, because the
agitation of the air which I produce will not affect
the strings. But that note will be heard quite well at
the other side of the screen. You must remember that
when the air carries a sound it vibrates at a certain rate

belonging to the sound. I make the air vibrate by singing a particular note, and if that rate of vibration corresponds to the strings the air will pass on part of its vibration to the strings, and so make the strings move. But if the rate of vibration is not the one that corresponds to the strings, then the air will not pass on any of its vibrations to the strings, and consequently the sound will be heard equally loud after it has passed through the strings. Having put the strings of the piano across the room, if I sing up the scale, when I come to the note which belongs to each of the strings my voice will suddenly appear to be deadened, because at the moment that the rate of vibration which I impress upon the air coincides with that belonging to the strings, part of it will be taken up in setting the strings in motion. As I pass the note, then, which belongs to the strings, that note will be deadened.

Instead of a screen of piano strings let us put in a series of sets of bells, three or four belonging to each set, so that each set of bells answers to three or four notes, and so that all the sets are exactly alike. Now suppose that these sets of bells are distributed all over the middle part of the room, and that I sing straight up the scale from one note to another until I come to the note that corresponds to one of the bells in these sets, then that note will appear to be deadened at the other end, because part of the vibration communicated to the air will be taken up in setting those bells in motion. When I come to another note which belongs to them, that note will also be deadened ; so that a person listening at the other end of the room would observe that certain notes were deadened, or even had disappeared

altogether. If, however, I sing loud enough, I then
should set all these bells vibrating. What would be
heard at the other end of the room? Why just the
chord compounded out of those sounds that belonged
to the bells, because the bells having been set vibrating
would give out the corresponding notes. So you see
there are here three facts. When I sing a note which
does not belong to the bells, my voice passes to the end
of the room without diminution. When I sing a note
that does belong to the bells, then if it is not loud enough
it is deadened by passing through the screen; but if it
is loud enough it sets the bells vibrating, and is heard
afterwards. Now just notice this consequence. We have
supposed a screen made out of these structures that I
have imagined to represent atoms, and when I sing
through the scale at one end of the room certain notes
appear to be deadened. If I take away half of those
structures, what will be the effect? Exactly the same
notes will be deadened, but they will not be deadened
so much; the notes which are picked out of the thinner
screen to be deadened will be exactly the same notes,
but the amount of the deadening will not be the same.

So far we have only been talking about the transmission of *sound*. You know that sound consists of
certain waves which are passed along in the air; they
are called 'aerial vibrations.' We also know that
light consists of certain waves which are passed along,
not in the air, but along another medium. I cannot
stop at present to explain to you what the sort of
evidence is upon which that assertion rests, but it is the
same sort of evidence as that which I shall try to show
you belongs to the statement about atoms; that is to

say, the 'undulatory theory,' as it is called, of light,
the theory that light consists of waves transmitted
along a certain medium, has passed out of the stage of
being a theory, and has passed into the stage of being
a demonstrated fact.   The difference between a theory
and a demonstrated fact is something like this: If you
supposed a man to have walked from Chorlton Town
Hall down here say in ten minutes, the natural conclusion would be that he had walked along the Stret-
ford Road.   Now that theory would entirely account
for all the facts, but at the same time the facts would
not be proved by it.   But suppose it happened to be
winter time, with snow on the road, and that you
could trace the man's footsteps all along the road, then
you would know that he had walked along that way.
The sort of evidence we have to show that light does
consist of waves transmitted through a medium is the
sort of evidence that footsteps upon the snow make;
it is not a theory merely which simply accounts for the
facts, but it is a theory which can be reasoned back to
from the facts without any other theory being possible.
So that you must just for the present take it for granted
that the arguments in favour of the hypothesis that
light consists of waves are such as to take it out of the
region of hypothesis, and make it into demonstrated fact.

Very well, then, light consists of waves transmitted
along this medium in the same way that sound is transmitted along the air.   The waves are not of the same
kind ; but still they are waves, and they are transmitted
as such ; and the different colours of light correspond
to the different lengths of these waves, or to the different
rates of the vibration of the medium, just as the different

pitches of sound correspond to the different lengths of
the air-waves or to the different rates of the vibration of
the air. Now, if we take any gas, such as oxygen, and we
pass light through it, we find that that gas intercepts,
or weakens, certain particular colours. If we take any
other gas, such as hydrogen, and pass light through it,
we find that that gas intercepts, or weakens, certain
other particular colours of the light. There are two
ways in which it can do that : it is clear that the un-
dulations, or waves, are made weaker, because they
happen to coincide with the rate of vibration of the
gas they are passing through. But the gas may vibrate
as a whole in the same way that the air does when you
transmit sound. Or the waves may be stopped, because
the gas consists of a number of small structures ; just
as my screen, which I imagine to consist of structures ;
or just as the screen of piano strings is made up of the
same structure many times repeated. Either of these
suppositions would apparently at first account for the
fact that certain waves of light are intercepted by
the gas, while others are let through. But how is it
that we can show one of these suppositions is wrong
and the other is right? Instead of taking so small a
structure as piano strings, let us suppose we had got
a series of fiddles, the strings of all of them being
stretched exactly in tune. I suppose this case because
it makes a more complicated structure, for there would
be two or three notes corresponding in each fiddle. If
you suppose this screen of fiddles to be hung up and
then compressed, what will be the effect? The effect
of the compression will be, if they are all in contact,
that each fiddle itself will be altered. If the fiddles

are compressed longways, the strings will give lower notes than before, and consequently the series of notes which will be intercepted by that screen will be different from the series of notes which were intercepted before. But if you have a screen made out of fiddles which are at a distance from one another, and then if you compress them into a smaller space by merely bringing them nearer together, without making them touch, then it is clear that exactly the same notes will be intercepted as before; only, as there will be more fiddles in the same space, the deadening of the sound will be greater.

Now when you compress any gas you find that it intercepts exactly the same colours of light which it intercepted before it was compressed. It follows, therefore, that the rates of vibration which it intercepts depend not upon the mass of the gas whose properties are altered by the compression, but upon some individual parts of it which were at a distance from one another before, and which are only brought nearer together without being absolutely brought into contact so as to squeeze them. That is the sort of reasoning by which it is made clear that the interception of light, or particular waves of light, by means of a gas, must depend on certain individual structures in the gas which are at a distance from one another, and which by compression are not themselves compressed, but only brought nearer to one another.

There is an extremely interesting consequence which follows from this reasoning, and which was deduced from it by Professor Stokes in the year 1851, and which was afterwards presented in a more developed form in the magnificent researches of Kirchhoff—namely, the reasoning about the presence of certain matter in the sun. If you analyse the solar light by passing it through a prism, the effect of the prism is to divide it off so as to separate the light into the different colours which it contains. That line of variously coloured light which is produced by the prism is, as you know, called the Spectrum. When that spectrum is made in a very accurate way, so that the parts of it are well defined, it is observed to contain certain dark lines. That is, there is a certain kind of light which is missing in the sunlight ; certain kinds of light, as we travel along the scale of lights, are missing. Why are they missing ? Because there is something that the light has passed through which intercepts or weakens those kinds of light. Now that something which the light has passed through, how shall we find out what it is ? It ought to be the same sort of substance which if it were heated would give out exactly that kind of light. Now there is a certain kind of light which is intercepted which makes a group of dark lines in the solar spectrum. There are two principal lines which together are called the line D ; and it is found that exactly that sort of light is emitted by sodium when heated hot enough. The conclusion therefore is that that matter which intercepts that particular part of the solar light is sodium, or that there is sodium somewhere between us and the hot portion of the sun which sends us the light. And other reasons lead us to conclude that this sodium is not in the atmosphere of the earth, but in the neighbourhood of the sun—that it exists in a gaseous state in the sun's atmosphere. And nearly all the lines in the solar spectrum have been explained in

that way, and shown to belong to certain substances
which we are able to heat here, and to show that when
they are heated they give out exactly the same kind of
light which they intercepted when the light was first
given out by the sun and they stood in the way So
you see that is a phenomenon exactly like the phenomenon presented by the finger-glass that we began
with.

Precisely the same light which any gas will give out
when it is heated, that same kind of light it will stop or
much weaken if the light is attempted to be passed
through it. That means that this medium which trans-
mits light, and which we call the ' luminiferous ether,'
has a certain rate of vibration for every particular colour
of the spectrum. When that rate of vibration coincides
with one of the rates of vibration of an atom, then it will
be stopped by that atom, because it will set the atom
vibrating itself. If therefore you pass light of any particular colour through a gas whose atoms are capable
of the corresponding rate of vibration, the light will be
cut off by the gas. If on the other hand you so far heat
the gas that the atoms are vibrating strongly enough to
give out light, it will give out a light of a kind which it
previously stopped.

We have reason then for believing that a simple gas
consists of a great number of atoms; that it consists of
very small portions, each of which has a complicated
structure, but that structure is the same for each of
them, and that these portions are separate, or that there
is space between them.

In the next place I want to show you what is the
evidence upon which we believe that these portions

of the gas are in motion—that they are constantly moving.

If this were a political instead of a scientific meeting, there would probably be some people who would be inclined to disagree with us, instead of all being inclined to agree with one another; and these people might have taken it into their heads, as has been done in certain cases, to stop the meeting by putting a bottle of sulphuretted hydrogen in one corner of the room and taking the cork out. You know that after a certain time the whole room would contain sulphuretted hydrogen, which is a very unpleasant thing to come in contact with. Now how is it that that gas which was contained in a small bottle could get in a short time over the whole room unless it was in motion? What we mean by motion is change of place. The gas was in one corner, and it is afterwards all over the room. There has therefore been motion somewhere, and this motion must have been of considerable rapidity, because we know that there was the air which filled the room beforehand to oppose resistance to that motion. We cannot suppose that the sulphuretted hydrogen gas was the only thing that was in motion, and that the air was not in motion itself, because if we had used any other gas we should find that it would diffuse itself in exactly the same way. An argument just like that applies also to the case of a liquid. Suppose this room were a large tank entirely filled with water and anybody were to drop a little iodine into it, after a certain time the whole of the water would be found to be tinged of a blue colour. Now that drop may be introduced into any part of the tank you like, either at the top or bottom,

and it will always diffuse itself over the whole water.
There has here again been motion. We cannot suppose
that the drop which was introduced was the only thing
that moved about, because any other substance would
equally have moved about. And the water has moved
into the place where the drop was, because in the place
where you put the drop there is not so much iodine as
there was to begin with. Well then it is clear that in
the case of a gas, these particles of which we have shown
it to consist must be constantly in motion ; and we have
shown also that a liquid must consist of parts that are in
motion, because it is able to admit the particles of another body among them.

When we have decided that the particles of a gas
are in motion, there are two things that they may do
—they may either hit against one another, or they may
not. Now it is established that they do hit against one
another, and that they do not proceed along straight
lines independent of one another. But I cannot at
present explain to you the whole of the reasoning upon
which that conclusion is grounded. It is grounded
upon some rather hard mathematics. It was shown by
Professor Clerk Maxwell that a gas cannot be a medium
consisting of small particles moved about in all directions
in straight lines, which do not interfere with one another,
but which bound off from the surfaces which contain
this medium. Supposing we had a box containing a
gas of this sort. Well, these particles do not interfere
with one another, but only rebound when they come
against the sides of the box ; then that portion of the
gas will behave not like a gas but like a solid body.
The peculiarity of liquids and gases is that they do not

mind being bent and having their shape altered.   It has been shown by Clerk Maxwell that a medium whose particles do not interfere with one another would behave like a solid body and object to be bent.   It was a most extraordinary conclusion to come to, but it is entirely borne out by the mathematical formulæ.   It is certain that if there were a medium composed of small particles flying about in all directions and not interfering with one another, then that medium would be to a certain extent solid, that is, would resist any bending or change of shape.   By that means then it is known that these particles do run against one another.   And they come apart again.   There were two things of course they might do : they might either go on in contact, or they might come apart.   Now we know that they come apart for this reason—we have already considered how two gases in contact will diffuse into one another.   If you were to put a bucket containing carbonic acid (which is very heavy) upon the floor of this room, it would after a certain time diffuse itself over all the room ; you would find carbonic acid gas in every part of the room. Graham found that if you were to cover over the top of that bucket with a very thin cover made out of graphite, or blacklead, then the gas would diffuse itself over the room pretty nearly as fast as before.   The graphite acts like a porous body, as a sponge does to water, and lets the gas get through.   The remarkable thing is that if the graphite is thin the gas will get through nearly as fast as it will if nothing is put between to stop it. Graham found out another fact.   Suppose that bucket to contain two very different gases, say a mixture of hydrogen and carbonic acid gas.   Then the hydrogen

would come out through the blacklead very much faster
than the carbonic acid gas. It is found by mathematical calculation that if you have two gases, which
are supposed to consist of small particles which are all
banging about, the gas whose particles are lightest will
come out quickest; that a gas which is four times as
light will come out twice as fast; and a gas nine times
as light will come out three times as fast, and so on.
Consequently, when you mix two gases together and
then pass them through a thin piece of blacklead, the
lightest gas comes out quickest, and is as it were sifted
from the other. Now suppose we put pure hydrogen
into a bucket and put blacklead on the top, and then see
how fast the hydrogen comes out. If the particles of
the hydrogen are different from one another, if some are
heavier, the lighter ones will come out first. Now let
us suppose we have got a vessel which is divided into
two parts by a thin wall of blacklead. We will put
hydrogen into one of these parts and allow it to come
through this blacklead into the other part; then if the
hydrogen contains any molecules or atoms which are
lighter than the others, those will come through first.
If we test the hydrogen that has come through, we shall
find that the atoms, as a rule, on one side of this wall
are lighter than the atoms on the other side. How
should we find that out? Why we should take these
two portions of gas, and we should try whether one of
them would pass through another piece of blacklead
quicker than the other; because if it did, it would
consist of lighter particles. Graham found that it did
not pass any quicker. Supposing you put hydrogen into
one half of such a vessel, and then allow the gas to diffuse

itself through the blacklead, the gas on the two sides would be found to be of precisely the same qualities. Consequently, there has not been in this case any sifting of the lighter particles from the heavier ones; and consequently there could not have been any lighter particles to sift, because we know that if there were any they would have come through quicker than the others. Therefore we are led to the conclusion that in any simple gas, such as hydrogen or oxygen, all the atoms are, as nearly as possible, of the same weight. We have no right to conclude that they are exactly of the same weight, because there is no experiment in the world that enables us to come to an exact conclusion of that sort. But we are enabled to conclude that, within the limits of experiment, all the atoms of a simple gas are of the same weight. What follows from that? It follows that when they bang against one another, they must come apart again; for if two of them were to go on as one, that one would be twice as heavy as the others, and would consequently be sifted back. It follows therefore that two particles of a gas which bang against one another must come apart again, because if they were to cling together they would form a particle twice as heavy, and so this clinging would show itself when the gas was passed through the screen of black-lead.

Now there are certain particles or small masses of matter which we know to bang against one another according to certain laws; such, for example, as billiard balls. The way in which different bodies, after hitting together, come apart again, depends on the constitution of those bodies. The earlier hypothesis about

the constitution of a gas supposed that the particles of
them came apart according to the same law that billiard
balls do; but that hypothesis, although it was found to
explain a great number of phenomena, did not explain
them all. And it was Professor Clerk Maxwell again
who found the hypothesis which does explain all the
rest of the phenomena. He found that particles when
they come together separate as if they repelled one
another, or pushed one another away; and as if they
did that much more strongly when close together than
when further apart. You know that what is called the
great law of gravitation asserts that all bodies pull one
another together according to a certain rule, and that
they pull one another more when close than when
further apart. Now that law differs from the law which
Clerk Maxwell found out as affecting the repulsion of
gaseous particles. The law of attraction of gravitation
is this; that when you halve the distance, you have to
multiply the attraction four times—twice two make
four. If you divide the distance into three, you must
multiply the attraction nine times—three times three
are nine. Now in the case of atomic repulsion you have
got to multiply not twice two, or three times three,
but five twos together—which multiplied make 32. If
you halve the distance between two particles you
increase the repulsion 32 times. So also five threes
multiplied together make 243; and if you divide the
distance between two particles by three, then you
increase the repulsion by 243. So you see the repulsion increases with enormous rapidity as the distance
diminishes. That law is expressed by saying that the
repulsion of two gases is inversely as the fifth power

of the distance. But I must warn you against supposing that that law is established in the same sense
that these other statements that we have been making
are established. That law is true provided that there
is a repulsion between two gaseous particles, and that
it varies as a power of the distance ; it is proved that if
there is any law of repulsion, and if the law is that it
varies as some power of the distance, then that power
cannot be any other than the fifth. It has not been
shown that the action between the two particles is not
something perhaps more complicated than this, but
which on the average produces the same results. But
still the statement that the action of gaseous molecules
upon one another can be entirely explained by the
assumption of a law like that, is the newest statement
in physics since the law of gravitation was discovered.
You know that there are other actions of matter which
apparently take place through intervening spaces and
which always follow the same law as gravitation, such
as the attraction or repulsion of magnetical or electrical
particles : those follow the same law as gravitation.
But here is a law of repulsion which follows a different
law from that of gravitation, and in that lies the extreme
interest of Professor Clerk Maxwell's investigation.

The next thing that I want to give you reasoning
for is again rather a hard thing in respect of the
reasoning, but the fact is an extremely simple and
beautiful one. It is this. Suppose I have two vessels,
say cylinders, with stoppers which do not fit upon the
top of the vessel, but slide up and down inside and yet
fit exactly. These two vessels are of exactly the same
size ; one of them contains hydrogen and the other con-

tains oxygen. They are to be of the same temperature
and pressure, that is to say they will bear exactly the
same weight on the top. Very well, these two vessels
having equal volumes of gas of the same pressure
and temperature will contain just the same number of
atoms in each, only the atoms of oxygen will be heavier
than the atoms of hydrogen. Now how is it that we
arrive at that result? I shall endeavour to explain the
process of reasoning. Boyle discovered a law about the
dependence of the pressure of a gas upon its volume
which showed that if you squeezed a gas into a smaller
space it will press so much the more as the space has
been diminished. If the space has been diminished
one-half, then the pressure is doubled ; if the space
is diminished to one-third, then the pressure is increased to three times what it was before. This
holds for a varying volume of the same gas. That
same law would tell us that if we put twice the quantity
of gas into the same space, we should get twice the
amount of pressure. Dalton made a new statement
of that law, which expresses it in this form, that when
you put more gas into a vessel which already contains
gas, the pressure that you get is the sum of the two
pressures which would be got from the two gases
separately. You will see directly that that is equivalent
to the other· law. But the importance of Dalton's
statement of the law is this, that it enabled the law to
be extended from the case of the same gas to the case
of two different gases. If instead of putting a pint
of oxygen into a vessel already containing a pint, I
were to put in a pint of nitrogen, I should equally get
a double pressure. The oxygen and nitrogen, when

mixed together, would exert the sum of the pressures upon the vessel that the oxygen and nitrogen would exert separately. Now the explanation of that pressure is this. The pressure of the gas upon the sides of the vessel is due to the impact of these small particles which are constantly flying about and impinging upon the sides of the vessel. It is first of all shown mathematically that the effect of that impinging would be the same as the pressure of the gas. But the amount of the pressure could be found if we knew how many particles there were in a given space, and what was the effect of each one when it impinged on the sides of the vessel. You see directly why it is that putting twice as many particles, which are going at the same rate, into the same vessel, we should get twice the effect. Although there are just twice as many particles to hit the sides of the vessel, they are apparently stopped by each other when they bound off. But the effect of there being more particles is to make them come back quicker ; so that altogether the number of impacts upon the sides of the vessel is just doubled when you double the number of particles. Supposing we have got a cubic inch of space, then the amount of pressure upon the side of that cubic inch depends upon the number of particles inside the cube, and upon the energy with which each one of them strikes against the sides of the vessel.

Again there is a law which connects together the pressure of a gas and its temperature. It is found that there is a certain absolute zero of temperature, and that if you reckon your temperature from that, then the pressure of the gas is directly proportional to the temperature, that twice the temperature will give twice the pressure of the same gas, and three times the temperature will give three times the pressure of the same gas.

Well now we have just got to remember these two rules—the law of Boyle, as expressed by Dalton, connecting together the pressure of a gas and its volume, and this law which connects together the pressure with the absolute temperature. You must remember that it has been calculated by mathematics that the pressure upon one side of a vessel of a cubic inch has been got by multiplying together the number of particles into the energy with which each of them strikes against the side of the vessel. If we keep that same gas in a vessel and alter its temperature, then we find that the pressure is proportional to the temperature; but since the number of molecules remains the same when we double the pressure, we must alter that other factor in the pressure, we must double the energy with which each of the particles attacks the side of the vessel. That is to say, when we double the temperature of the gas we double the energy of each particle; consequently the temperature of the gas is proportional always to the energy of its particles. That is the case with a single gas. If we mix two gases, what happens? They come to exactly the same temperature. It is calculated also by mathematics that the particles of one gas have the same effect as those of the other; that is, the light particles go faster to make up for their want of weight. If you mix oxygen and hydrogen, you find that the particles of hydrogen go four times as fast as the particles of oxygen. Now we have here a mathematical statement

—that when two gases are mixed together, the energy of the two particles is the same ; and with any one gas considered by itself that energy is proportional to the temperature. Also when two gases are mixed together the two temperatures become equal. If you think over that a little, you will see that it proves that whether we take the same gas or different gases, the energy of the single particles is always proportional to the temperature of the gas.

What follows? If I have two vessels containing gas at the same pressure and the same temperature (suppose that hydrogen is in one and oxygen in the other), then I know that the temperature of the hydrogen is the same as the temperature of the oxygen, and that the pressure of the hydrogen is the same as the pressure of the oxygen. I also know (because the temperatures are equal) that the average energy of a particle of the hydrogen is the same as that of a particle of the oxygen. Now the pressure is made up by multiplying the energy by the number of particles in both gases ; and as the pressure in both cases is the same, therefore the number of particles is the same. That is the reasoning ; I am afraid it will seem rather complicated at first hearing, but it is this sort of reasoning which establishes the fact that in two equal volumes of different gases at the same temperature and pressure, the number of particles is the same.

Now there is an exceedingly interesting conclusion which was arrived at very early in the theory of gases, and calculated by Mr. Joule. It is found that the pressure of a gas upon the sides of a vessel may be represented quite fairly in this way. Let us divide the particles of gas into three companies or bands. Suppose I have a cubical vessel in which one of these companies is to go forward and backward, another right and left, and the other to go up and down. If we make those three companies of particles to go in their several directions, then the effect upon the sides of the vessel will not be altered; there will be the same impact and pressure. It was also found out that the effect of this pressure would not be altered if we combined together all the particles forming one company into one mass, and made them impinge with the same velocity upon the sides of the vessel. The effect of the pressure would be just the same. Now we know what the weight of a gas is, and we know what the pressure is that it produces, and we want to find the velocity it is moving at on the average. We can find out at what velocity a certain weight has to move in order to produce a certain definite impact. Therefore we have merely to take the weight of the gas, divide it by three, and to find how fast that has to move in order to produce the pressure, and that will give us the average rate at which the gas is moving. By that means Mr. Joule calculated that in air of ordinary temperature and pressure the velocity is about 500 mètres per second, nearly five miles in sixteen seconds, or nearly twenty miles a minute— about sixty times the rate of an ordinary train.

The average velocity of the particles of gas is about $1\frac{1}{2}$ times as great as the velocity of sound. You can easily remember the velocity of sound in air at freezing point—it is 333 metres per second; so that about $1\frac{1}{2}$ times, really 1 432 of that would be the average velocity of a particle of air. At the ordinary temperature—60 degrees Fahrenheit—the velocity would, of course, be greater.

Let us consider how much we have established so far about these small particles of which we find that the gas consists. We have so far been treating mainly of gases. We find that a gas, such as the air in this room, consists of small particles, which are separate with spaces between them. They are as a matter of fact of two different types, oxygen and nitrogen. All the particles of oxygen contain the same structure, and the rates of internal vibration are the same for all these particles. It is also compounded of particles of nitrogen which have different rates of internal vibration. We have shown that these particles are moving about constantly. We have shown that they impinge against and interfere with one another's motion; and we have shown that they come apart again. We have shown that in vessels of the same size containing two different gases of the same pressure and temperature there is the same number of those two different sorts of particles. We have shown also that the average velocity of these particles in the air of this room is about twenty miles a minute.

There is one other point of very great interest to which I want to call your attention. The word 'atom,' as you know, has a Greek origin; it means that which is not divided. Various people have given it the meaning of that which *cannot* be divided; but if there is anything which cannot be divided we do not know it, because we know nothing about possibilities or impossibilities, only about what has or has not taken

place.   Let us then take the word in the sense in which
it can be applied to a scientific investigation.   An atom
means something which is not divided *in certain cases
that we are considering*.   Now these atoms I have been
talking about may be called physical atoms, because
they are not divided under those circumstances that
are considered in physics.   These atoms are not divided
under the ordinary alteration of temperature and
pressure of gas, and variation of heat; they are not in
general divided by the application of electricity to the
gas, unless the stream is very strong.   But there is a
science which deals with operations by which these
atoms which we have been considering can be divided
into two parts, and in which therefore they are no longer
atoms.   That science is chemistry.   The chemist therefore will not consent to call these little particles that we
are speaking of by the name of atoms, because he knows
that there are certain processes to which he can subject
them which will divide them into parts, and then
they cease to be things which have not been divided.
I will give you an instance of that.   The atoms of
oxygen which exist in enormous numbers in this room
consist of two portions, which are of exactly the same
structure.   Every molecule, as the chemist would call
it, travelling in this room, is made up of two portions
which are exactly alike in their structure.   It is a
complicated structure; but that structure is double.
It is like the human body—one side is like the other
side.   How do we know that?   We know it in this way.
Suppose that I take a vessel which is divided into two
parts by a division which I can take away.   One of
these parts is twice as large as the other part, and will

contain twice as much gas. Into that part which is twice as big as the other I put hydrogen; into the other I put oxygen. Suppose that one contains a quart and the other a pint; then I have a quart of hydrogen and a pint of oxygen in this vessel. Now I will take away the division so that they can permeate one another, and then if the vessel is strong enough I pass an electric spark through them. The result will be an explosion inside the vessel; it will not break if it is strong enough; but the quart of hydrogen and the pint of oxygen will be converted into steam; they will combine together to form steam. If I choose to cool down that steam until it is just as hot as the two gases were before I passed the electric spark through them, then I shall find that at the same pressure there will only be a quart of steam. Now let us remember what it was that we established about two equal volumes of different gases at the same temperature and pressure. First of all, we had a quart of hydrogen with a pint of oxygen. We know that that quart of hydrogen contains twice as many hydrogen molecules as the pint of oxygen contains of oxygen molecules. Let us take particular numbers. Suppose instead of a quart or a pint we take a smaller quantity, and say that there are 100 hydrogen and 50 oxygen molecules. Well, after the cooling has taken place, I should find a volume of steam which was equal to the volume of hydrogen, that is, I should find 100 steam molecules. Now these steam molecules are made up of hydrogen and oxygen molecules. I have got therefore 100 things which are all exactly alike, made up of 100 things and 50 things—100 hydrogen and 50 oxygen, making 100 steam molecules. Now since the 100 steam molecules

are exactly alike, we have those 50 oxygen molecules
distributed over the whole of these steam molecules.
Therefore, unless the oxygen contains something which
is common to the hydrogen also, it is clear that each of
those 50 molecules of oxygen must have been divided
into two, because you cannot put 50 horses into 100
stables, so that there shall be exactly the same amount
of horse in each stable; but you can divide 50 *pairs* of
horses among 100 stables.   There we have the supposition that there is nothing common to the oxygen and
hydrogen, that there is no structure that belongs to each
of them.   Now that supposition is made by a great
majority of chemists.   Sir Benjamin Brodie, however, has
made a supposition that there is a structure in hydrogen which is also common to certain other elements.
He has himself, for particular reasons, restricted that
supposition to the belief that hydrogen is contained as a
whole in many of the other elements.   Let us make that
further supposition and it will not alter our case at all.
We have then one hundred hydrogen and fifty oxygen
molecules, but there is something common to the two.
Well, this something we will call X.   Of this we have to
make one hundred equal portions.   Now that cannot be
the case unless that structure occurred twice as often in
each molecule of oxygen as in each molecule of hydrogen.
Consequently, whether the oxygen molecule contains
something common to hydrogen or not, it is equally true
that the oxygen molecule must contain the same thing
repeated twice over ; it must be divisible into two parts
which are exactly alike.

Similar reasoning applies to a great number of other
elements ; to all those which are said to have an even

number of atomicities. But with regard to those which are said to have an odd number, although many of these also are supposed to be double, yet the evidence in favour of that supposition is of a different kind; and we must regard the supposition as still a theory and not yet a demonstrated fact.

Now I have spoken so far only of gases. I must for one or two moments refer to some calculations of Sir William Thomson, which are of exceeding interest as showing us what is the proximity of the molecules in liquids and in solids. By four different modes of argument derived from different parts of science, and pointing mainly to the same conclusion, he has shown that the distance between two molecules in a drop of water is such that there are between five hundred millions and five thousand millions of them in an inch. He expresses that result in this way—that if you were to magnify a drop of water to the size of the earth, then the coarseness of the graining of it would be something between that of cricket-balls and small shot. Or we may express it in this rather striking way. You know that the best microscopes can be made to magnify from 6,000 to 8,000 times. A microscope which would magnify that result as much again would show the molecular structure of water.

There is another scientific theory analogous to this one which leads us to hope that some time we shall know more about these molecules. You know that since the time that we have known all about the motions of the solar system, people have speculated about the origin of it; and a theory started by Laplace and worked out by other people has, like the theory of luminiferous ether, been taken out of the rank of hypothesis into that of fact. We know the rough outlines of the history of the solar system, and there are hopes that when we know the structure and properties of a molecule, what its internal motions are and what are the parts and shape of it, somebody may be able to form a theory as to how that was built up and what it was built out of. It is obvious that until we know the shape and structure of it, nobody will be able to form such a theory. But we can look forward to the time when the structure and motions in the inside of a molecule will be so well known that some future Kant or Laplace will be able to make an hypothesis about the history and formation of matter.[1]

[1] The mathematical development of this subject is due to Clausius and Maxwell. Reference to the chief papers will be found at the beginning of Maxwell's Memoir, 'On the Dynamical Theory of Gases,' Phil. Trans., 1867.

THE FIRST AND THE LAST CATASTROPHE.[1]

A CRITICISM ON SOME RECENT SPECULATIONS ABOUT THE DURATION OF THE UNIVERSE.

I PROPOSE in this lecture to consider speculations of quite recent days about the beginning and the end of the world. The world is a very interesting thing, and I suppose that from the earliest times that men began to form any coherent idea of it at all, they began to guess in some way or other how it was that it all began, and how it was all going to end. But there is one peculiarity about these speculations which I wish now to consider, that makes them quite different from the early guesses of which we read in many ancient books. These modern speculations are attempts to find out how things began, and how they are to end, by consideration of the way in which they are going on now. And it is just that character of these speculations that gives them their interest for you and for me; for we have only to consider these questions from the scientific point of view. By the scientific point of view I mean one which attempts to apply past experience to new circumstances according to an observed order of nature. So that we shall only consider the way in which things began, and the way in which they are to end, in so far as we seem

[1] Sunday Lecture Society, April 12, 1874: afterwards revised for publication.

able to draw inferences about the questions from facts which we know about the way in which things are going on now. And, in fact, the great interest of the subject to me lies in the amount of illustration which it offers of the degree of knowledge which we have now attained of the way in which the universe is going on.

The first of these speculations is one set forth by Professor Clerk Maxwell, in a lecture on Molecules delivered before the British Association at Bradford. Now, this argument of his which he put before the British Association at Bradford depends entirely upon the modern theory of the molecular constitution of matter. I think this the more important, because a great number of people appear to have been led to the conclusion that this theory is very similar to the guesses which we find in ancient writers —Democritus and Lucretius. It so happens that these ancient writers did hold a view of the constitution of things which in many striking respects agrees with the view which we hold in modern times. This parallelism has been brought recently before the public by Professor Tyndall in his excellent address at Belfast. And it is perhaps on account of the parallelism, which he pointed out at that place, between the theories held amongst the ancients and the theory held amongst the moderns, that many people who are acquainted with classic literature have thought that a knowledge of the views of Democritus and Lucretius would enable them to understand and criticize the modern theory of matter. That, however, is a mistake. The difference between the two is mainly this : the atomic theory of Democritus was a guess, and no more than a guess. Everybody around him was guessing about the origin of things, and they guessed in a great number of ways; but he happened to make a guess which was more near the right thing than any of the others. This view was right in its main hypothesis—that all things are made up of elementary parts, and that the different properties of different things depend rather upon difference of arrangement than upon ultimate difference in the substance of which they are composed. Although this was contained in the atomic theory of Democritus, as expounded by Lucretius, yet it will be found by anyone who examines further the consequences which are drawn from it that it very soon diverges from the truth of things, as we might naturally expect it would. On the contrary, the view of the constitution of matter which is held by scientific men in the present day is not a guess at all.

In the first place I will endeavour to explain what are the main points in this theory. First of all we must take the simplest form of matter, which turns out to be a gas—such, for example, as the air in this room. The belief of scientific men in the present day is that this air is not a continuous thing, that it does not fill the whole of the space in the room, but is made up of an enormous number of exceedingly small particles. There are two sorts of particles: one sort of particle is oxygen, and another sort of particle nitrogen. All the particles of oxygen are as near as possible alike in these two respects; first in weight, and secondly in certain peculiarities of mechanical structure. These small molecules are not at rest in the room, but are flying about in all directions with a mean velocity of seventeen miles a

minute. They do not fly far in one direction; but any
particular molecule, after going over an incredibly
short distance—the measure of which has been made—
meets another, not exactly plump, but a little on one
side, so that they behave to one another somewhat in
the same way as two people do who are dancing Sir Roger
de Coverley; they join hands, swing round, and then fly
away in different directions. All these molecules are
constantly changing the direction of each other's motion;
they are flying about with very different velocities,
although, as I have said, their mean velocity is about
seventeen miles a minute. If the velocities were all
marked off on a scale, they would be found distributed
about the mean velocity just as shots are distributed
about a mark. If a great many shots are fired at a
target, the hits will be found thickest at the bull's-eye,
and they will gradually diminish as we go away from that,
according to a certain law which is called the law of
error. It was first stated clearly by Laplace; and it
is one of the most remarkable consequences of theory
that the molecules of a gas have their velocities distributed amongst them precisely according to this law of
error. In the case of a liquid, it is believed that the
state of things is quite different. We said that in the
gas the molecules are moved in straight lines, and that
it is only during a small portion of their motion that
they are deflected by other molecules; but in a liquid
we may say that the molecules go about as if they were
dancing the grand chain in the Lancers. Every molecule after parting company with one finds another, and
so is constantly going about in a curved path, and never
sent quite clear away from the sphere of action of the

surrounding molecules. But, notwithstanding that, all molecules in a liquid are constantly changing their places, and it is for that reason that diffusion takes place in the liquid. Take a large tank of water and drop a little iodine into it, and you will find after a certain time all the water turned slightly blue. That is because all the iodine molecules have changed like the others and spread themselves over the whole of the tank. Because, however, you cannot see this, except where you use different colours, you must not suppose that it does not take place where the colours are the same. In every liquid all the molecules are running about and continually changing and mixing themselves up in fresh forms. In the case of a solid quite a different thing takes place. In a solid every molecule has a place which it keeps; that is to say, it is not at rest any more than a molecule of a liquid or a gas, but it has a certain mean position which it is always vibrating about and keeping fairly near to, and it is kept from losing that position by the action of the surrounding molecules. These are the main points of the theory of the constitution of matter as at present believed.

It differs from the theory of Democritus in this way. There is no doubt that in the first origin of it, when it was suggested to the mind of Daniel Bernouilli as an explanation of the pressure of gases, and to the mind of Dalton as an explanation of chemical reactions, it was a guess; that is to say, it was a supposition which would explain these facts of physics and chemistry, but which was not known to be true. Some theories are still in that position; other theories are known to be true, because they can be argued back to from the facts. In order to make out that your supposition is true, it is necessary to show, not merely that that particular supposition will explain the facts, but also that no other one will. Now, by the efforts of Clausius and Clerk Maxwell, the molecular theory of matter has been put in this other position. Namely, instead of saying, Let us suppose such and such things are true,—and then deducing from that supposition what the consequences ought to be, and showing that these consequences are just the facts which we observe—instead of doing that, I say, we make certain experiments ; we show that certain facts are undoubtedly true, and from these facts we go back by a direct chain of logical reasoning, which there is no way of getting out of, to the statement that all matter is made up of separate pieces or molecules, and that in matter of a given kind, in oxygen, or in hydrogen, or in nitrogen, these molecules are of very nearly the same weight, and have certain mechanical properties which are common to all of them. In order to show you something of the kind of evidence for that statement, I must mention another theory which, as it seems to me, is in the same position ; namely, the doctrine of the luminiferous ether, or that wonderful substance which is distributed all over space, and which carries light and radiant heat. By means of certain experiments upon interference of light, we can show, not by any hypothesis, not by any guess at all, but by a pure interpretation of the experiment—that in every ray of light there is some change or other, whatever it is, which is periodic in time and in space. By saying it is periodic in time, I mean that, at a given point of the ray of light, this change increases up to a certain instant, then decreases, then increases in the opposite direction, and then decreases again, and so on alternately. That is shown by experiments of interference; it is not a theory which will explain the facts, but it is a fact which is got out of observation. By saying that this phenomenon is periodic in space, I mean that, if at any given instant you could examine the ray of light, you would find that some change or disturbance, whatever it is, has taken place all along it in different degrees. It vanishes at certain points, and between these it increases gradually to a maximum on one side and the other alternately. That is to say, in travelling along a ray of light there is a certain change (which can be observed by experiments, by operating upon a ray of light with other rays of light), which goes through a periodic variation in amount. The height of the sea, as you know if you travel along it, goes through certain periodic changes ; it increases and decreases, and increases and decreases again at definite intervals. And if you take the case of waves travelling over the sea, and place yourself at a given point, or mark a point by putting a cork upon the surface, you will find that the cork will rise up and down ; that is to say, there will be a change or displacement of the cork's position, which is periodic in time, which increases and decreases, then increases in the opposite direction, and decreases again. Now this fact, which is established by experiment, and which is not a guess at all—the fact that light is a phenomenon periodic in time and space—is what we call the wave theory of light. The word ' theory ' here does not mean a guess ; it means an organized account of the facts, such that from it you may deduce results which are applicable to future experiments, the like of which have not yet been made. But we can see more than this. So far we say that light consists of waves, merely in the sense that it consists of some phenomenon or other which is periodic in time and in place; but we know that a ray of light or heat is capable of doing work. Radiant heat, for example, striking on a body, will warm it and enable it to do work by expansion; therefore this periodic phenomenon which takes place in the ray of light is something or other which possesses mechanical energy, which is capable of doing work. We may make it, if you like, a mere matter of definition, and say: Any change which possesses energy is a motion of matter; and this is perhaps the most intelligible definition of matter that we can frame. In that sense, and in that sense only, it is a matter of demonstration, and not a matter of guess, that light consists of the periodic motion of matter, of something which is between the luminous object and our eyes.

But that something is not matter in the ordinary sense of the term; it is not made up of such molecules as gases and liquids and solids are made up of. This last statement again is no guess, but a proved fact. There are people who ask: Why is it necessary to suppose a luminiferous ether to be anything else except molecules of matter in space, in order to carry light about? The answer is a very simple one. In order that separate molecules may carry about a disturbance, it is necessary that they should travel at least as fast as the disturbance travels. Now we know, by means that I shall afterwards come to, that the molecules of gas travel at a very ordinary rate—about twenty times as

fast as a good train.   But, on the contrary, we know by
the most certain of all evidence, by five or six different
means, that the velocity of light is 200,000 miles a
second.   By that very simple consideration we are able
to tell that it is quite impossible for light to be carried
by the molecules of ordinary matter, and that it wants
something else that lies between those molecules to
carry the light.   Now, remembering the evidence which
we have for the existence of this ether, let us consider
another piece of evidence ; let us now consider what
evidence we have that the molecules of a gas are separate from one another and have something between
them.   We find out, by experiment again, that the
different colours of light depend upon the various
rapidity of these waves, depend upon the size and upon
the length of the waves that travel through the ether,
and that when we send light through glass or any transparent medium except a vacuum, the waves of different
lengths travel with different velocities.   That is the case
with the sea ; we find that long waves travel faster than
short ones.   In much the same way, when light comes
out of a vacuum and impinges upon any transparent
medium, say upon glass, we find that the rate of transmission of all the light is diminished ; that it goes slower
when it gets inside of a material body ; and that this
change is greater in the case of small waves than of large
ones.   The small waves correspond to blue light, and
the large waves correspond to red light.   The waves of
red light are not made to travel so slowly as the waves
of blue light ; but, as in the case of waves travelling over
the sea, when light moves in the interior of a transparent
body the largest waves travel most quickly.   Well, then,

by using such a body as will separate out the different colours—a prism—we are able to affirm what are the constituents of the light which strikes upon it. The light that comes from the sun is made up of waves of various lengths; but, making it pass through a prism, we can separate it out into a spectrum, and in that way we find a band of light instead of a spot coming from the sun, and to every band in the spectrum corresponds a wave of a certain definite length and definite time in vibration. Now we come to a very singular phenomenon. If you take a gas such as chlorine and interpose it in the path of that light, you will find that certain particular rays of the spectrum are absorbed, while others are not. How is it that certain particular rates of vibration can be absorbed by this chlorine gas, while others are not? That happens in this way—that the chlorine gas consists of a great number of very small structures, each of which is capable of vibrating internally. Each of these structures is complicated, and is capable of a change of relative position amongst its parts of a vibratory character. We know that molecules are capable of such internal vibrations—for this reason, that if we heat any solid body sufficiently it will in time give out light; that is to say, the molecules are got into such a state of vibration that they start the ether vibrating, and they start the ether vibrating at the same rate at which they vibrate themselves. So that what we learn from the absorption of certain particular rays of light by chlorine gas is that the molecules of that gas are structures which have certain natural rates of vibration which they absorb, precisely those rates of vibration which belong to the molecules naturally. If you sing a certain note to a string of a piano, that string if in tune will vibrate. If, therefore, a screen of such strings were put across a room, and you sang a note on one side, a person on the other side would hear the note very weakly or not at all, because it would be absorbed by the strings; but if you sang another note, not one to which the strings naturally vibrated, then it would pass through, and would not be eaten up by setting the strings vibrating. Now this question arises. Let us put the molecules aside for a moment. Suppose we do not know of their existence, and say: Is this rate of vibration which naturally belongs to the gas a thing which belongs to it as a whole, or does it belong to the separate parts of it? You might suppose that it belongs to the gas as a whole. A jar of water, if you shake it, has a perfectly definite time in which it oscillates, and that is very easily measured. That time of oscillation belongs to the jar of water as a whole. It depends upon the weight of the water and the shape of the jar. But now, by a very certain method, we know that the time of vibration which corresponds to a certain definite gas does not belong to it as a whole, but belongs to the separate parts of it—for this reason, that if you squeeze the gas you do not alter the time of vibration. Let us suppose that we have a great number of fiddles in a room which are all in contact, and have strings accurately tuned to vibrate to certain notes. If you sang one of those notes all the fiddles would answer; but if you compress them you clearly put them all out of tune. They are all in contact, and they will not answer to the note with the same precision as before. But if you have a room which is full of fiddles, placed

at a certain distance from one another, then if you
bring them within shorter distances of one another, so
that they still do not touch, they will not be put out of
tune—they will answer exactly to the same note as
before. We see, therefore, that since compression of a
gas within certain limits does not alter the rate of vibration which belongs to it, that rate of vibration cannot
belong to the body of gas as a whole, but it must belong
to the individual parts of it. Now, by such reasoning
as this it seems to me that the modern theory of the constitution of matter is put upon a basis which is absolutely
independent of hypothesis. The theory is simply an
organized statement of the facts ; a statement, that is,
which is rather different from the experiments, being
made out from them in just such a way as to be most
convenient for finding out from them what will be the
results of other experiments. That is all we mean at
present by scientific theory.

Upon this theory Professor Clerk Maxwell founded
a certain argument in his lecture before the British
Association at Bradford. It is a consequence of the
molecular theory, as I said before, that all the molecules
of a certain given substance, say oxygen, are as near as
possible alike in two respects—first in weight, and
secondly in their times of vibration. Professor Clerk
Maxwell's argument was this. He first of all said
that the theory required us to believe not that these
molecules were as near as may be alike, but that they
were exactly alike in these two respects—at least the
argument appeared to me to require that. Then he said
all the oxygen we know of, whatever processes it has
gone through—whether it is got out of the atmosphere,

or out of some oxide of iron or carbon, or whether it belongs to the sun or the fixed stars, or the planets or the nebulæ—all this oxygen is alike. And all these molecules of oxygen we find upon the earth must have existed unaltered, or appreciably unaltered, during the whole of the time the earth has been evolved. Whatever vicissitudes they have gone through, however many times they have entered into combination with iron or carbon and been carried down beneath the crust of the earth, or set free and sent up again through the atmosphere, they have remained steadfast to their original form unaltered, the monuments of what they were when the world began. Professor Clerk Maxwell argues that things which are unalterable, and are exactly alike, cannot have been formed by any natural process. Moreover, being exactly alike, they cannot have existed for ever, and therefore they must have been made. As Sir John Herschel said, ' they bear the stamp of the manufactured article.'

Into these further deductions I do not propose to enter at all. I confine myself strictly to the first of the deductions which Professor Clerk Maxwell made from the molecular theory. He said that because these molecules are exactly alike, and because they have not been in the least altered since the beginning of time, therefore they cannot have been produced by any process of evolution. It is just that question which I want to discuss. I want to consider whether the evidence we have to prove that these molecules are exactly alike is sufficient to make it impossible that they can have been produced by any process of evolution.

The position that this evidence is not sufficient is

evidently by far the easier to defend; because the
negative is proverbially hard to prove; and if anyone
should prove that a process of evolution was impossible,
it would be an entirely unique thing in science and
philosophy.   In fact, we may see from this example precisely how great is the influence of authority in matters
of science.   If there is any name among contemporary
natural philosophers to whom is due the reverence of
all true students of science, it is that of Professor Clerk
Maxwell.   But if any one not possessing his great
authority had put forward an argument, founded
apparently upon a scientific basis, in which there
occurred assumptions about what things can and what
things cannot have existed from eternity, and about the
exact similarity of two or more things established by
experiment, we should say: 'Past eternity; absolute
exactness; this won't do;' and we should pass on to
another book.   The experience of all scientific culture
for all ages during which it has been a light to men
has shown us that we never do get at any conclusions
of that sort.   We do not get at conclusions about
infinite time or infinite exactness.   We get at conclusions
which are as nearly true as experiment can show, and
sometimes which are a great deal more correct than
direct experiment can be, so that we are able actually
to correct one experiment by deductions from another;
but we never get at conclusions which we have a right
to say are absolutely exact; so that even if we find a
man of the highest powers saying that he had reason to
believe a certain statement to be exactly true, or that
he believed a certain thing to have existed from the
beginning exactly as it is now, we must say: 'It is quite

possible that a man of so great eminence may have found out something which is entirely different from the whole of our previous knowledge, and the thing must be inquired into. But, notwithstanding that, it remains a fact that this piece of knowledge will be absolutely of a different kind from anything that we knew before.'

Now let us examine the evidence by which we know that the molecules of the same gas are as near as may be alike in weight and in rates of vibration. There were experiments made by Dr. Graham, late Master of the Mint, upon the rate at which different gases were mixed together. He found that if he divided a vessel by a thin partition made of black-lead or graphite, and put different gases on the two opposite sides, they would mix together nearly as fast as though there was nothing between them. The difference was that the plate of graphite made it more easy to measure the rate of mixture ; and Dr. Graham made measurements and came to conclusions which are exactly such as are required by the molecular theory. It is found by a process of mathematical calculation that the rate of diffusion of different gases depends upon the weight of the molecules. A molecule of oxygen is sixteen times as heavy as a molecule of hydrogen, and it is found upon experiment that hydrogen goes through a septum or wall of graphite four times as fast as oxygen does. Four times four are sixteen. We express that rule in mathematics by saying that the rate of diffusion of gas is inversely as the square root of the mass of its molecules. If one molecule is thirty-six times as heavy as another —the molecule of chlorine is nearly that multiple of hydrogen—it will diffuse itself at one-sixth of the rate.

This rule is a deduction from the molecular theory, and it is found, like innumerable other such deductions, to come right in practice. But now observe what is the consequence of this. Suppose that, instead of taking one gas and making it diffuse itself through a wall, we take a mixture of two gases. Suppose we put oxygen and hydrogen into one side of a vessel which is divided into two parts by a wall of graphite, and we exhaust the air from the other side, then the hydrogen will go through this wall four times as fast as the oxygen will. Consequently, as soon as the other side is full there will be a great deal more hydrogen in it than oxygen—that is to say, we shall have sifted the oxygen from the hydrogen, not completely, but in a great measure, precisely as by means of a screen we can sift large coals from small ones. Now let us suppose that when we have oxygen gas unmixed with any other the molecules are of two sorts and of two different weights. Then you see that if we make that gas pass through a porous wall, the lighter particles would pass through first, and we should get two different specimens of oxygen gas, in one of which the molecules would be lighter than in the other. The properties of one of these specimens of oxygen gas would necessarily be different from those of the other, and that difference might be found by very easy processes. If there were any perceptible difference between the average weight of the molecules on the two sides of the septum, there would be no difficulty in finding that out. No such difference has ever been observed. If we put any single gas into a vessel, and we filter it through a septum of black-lead into another vessel, we find no difference between the gas on one side of the wall and the gas on the other side. That is to say, if there is any difference it is too small to be perceived by our present means of observation. It is upon that sort of evidence that the statement rests that the molecules of a given gas are all very nearly of the same weight. Why do I say *very nearly* ? Because evidence of that sort can never prove that they are exactly of the same weight. The means of measurement we have may be exceedingly correct, but a certain limit must always be allowed for deviation ; and if the deviation of molecules of oxygen from a certain standard of weight were very small, and restricted within small limits, it would be quite possible for our experiments to give us the results which they do now. Suppose, for example, the variation in the size of the oxygen atoms were as great as that in the weight of different men, then it would be very difficult indeed to tell by such a process of sifting what that difference was, or in fact to establish that it existed at all. But, on the other hand, if we suppose the forces which originally caused all those molecules to be so nearly alike as they are to be constantly acting and setting the thing right as soon as by any sort of experiment we set it wrong, then the small oxygen atoms on one side would be made up to their right size, and it would be impossible to test the difference by any experiment which was not quicker than the processes by which they were made right again.

There is another reason why we are obliged to regard that experiment as only approximate, and as not giving us any exact results. There is very strong evidence, although it is not conclusive, that in a given gas—say in a vessel full of carbonic acid—the molecules are not all of the same weight. If we compress the gas, we find that when in the state of a perfect gas, or nearly so, the pressure increases just in the ratio that the volume diminishes. That law is entirely explained by means of the molecular theory. It is what ought to exist if the molecular theory is true. If we compress the gas further, we find that the pressure is smaller than it ought to be according to this law. This can be explained in two ways. First of all we may suppose that the molecules are so crowded that the time during which they are sufficiently near to attract each other sensibly becomes too large a proportion of the whole time to be neglected; and this will account for the change in the law. There is, however, another explanation. We may suppose, for illustration, that two molecules approach one another, and that the speed at which one is going relatively to the other is very small, and then that they so direct one another that they get caught together, and go on circling, making only one molecule. This, on scientific principles, will account for our fact, that the pressure in a gas which is near a liquid state is too small—that instead of the molecules going about singly, some are hung together in couples and some in larger numbers, and making still larger molecules. This supposition is confirmed very strikingly by the spectroscope. If we take the case of chlorine gas, we find that it changes colour—that it gets darker as it approaches the liquid condition. This change of colour means that there is a change in the rate of vibration which belongs to its component parts; and it is a very simple mechanical deduction that the larger molecules will, as a rule, have a slower rate of vibration than the smaller ones— very much in the same way as a short string gives a higher note than a long one. The colour of chlorine changes just in the way we should expect if the molecules, instead of going about separately, were hanging together in couples; and the same thing is true of a great number of the metals. Mr. Lockyer, in his admirable researches, has shown that several of the metals and metalloids have various spectra, according to the temperature and the pressure to which they are exposed; and he has made it exceedingly probable that these various spectra—that is, the rates of vibration of the molecules—depend upon the molecules being actually of different sizes. Dr. Roscoe has a few months ago shown an entirely new spectrum of the metal sodium, whereby it appears that this metal exists in a gaseous state in four different degrees of aggregation—as a simple molecule, and as three or four or eight molecules together. Every increase in the complication of the molecules—every extra molecule you hang on to the aggregate that goes about together—will make a difference in the rate of the vibration of that system, and so will make a difference in the colour of the substance.

So then we have an evidence of an entirely extraneous character that in a given gas the actual molecules that exist are not all of the same weight. Any experiment which failed to detect this would fail to detect any smaller difference. And here also we can see a reason why, although a difference in the size of

the molecules does exist, yet we do not find that out
by sifting. Suppose you take oxygen gas consisting of
single molecules and double molecules, and you sift it
through a plate ; the single molecules get through first,
but, when they get through, some of them join themselves together as double molecules ; and although more
double molecules are left on the other side, yet some
of them break up and make single molecules ; so the
process of sifting, which ought to give you single
molecules on the one side and double on the other,
merely gives you a mixture of single and double on both
sides ; because the reasons which originally decided that
there should be just those two forms are always at work
and continually setting things right.

Now let us take the other point in which molecules
are very nearly alike ; namely that they have very nearly
the same rate of vibration. The metal sodium in the
common salt upon the earth has two rates of vibration ;
it sounds two notes, as it were, which are very near to
each other. They form the well-known double line D
in the yellow part of the spectrum. These two bright
yellow lines are very easy to observe. They occur in
the spectra of a great number of stars. They occur in
the solar spectrum as dark lines, showing that there is
sodium in the outer rim of the sun, which is stopping
and shutting off the light of the bright parts behind.
All these lines of sodium are just in the same position in
the spectrum, showing that the rates of vibration of all
these molecules of sodium all over the universe, so far
as we know, are as near as possible alike. That implies
a similarity of molecular structure, which is a great deal
more delicate than any mere test of weight. You may

weigh two fiddles until you are tired, and you will never
find out whether they are in tune; the one test is a
great deal more delicate than the other.   Let us see how
delicate this test is.   Lord Rayleigh has remarked that
there is a natural limit for the precise position of a given
line in the spectrum, and for this reason.   If a body
which is emitting a sound comes towards you, you will
find that the pitch of the sound is altered.   Suppose
that omnibuses run every ten minutes in the streets, and
you walk in a direction opposite to that in which they
are coming, you will obviously pass more omnibuses in
an hour than if you walked in an opposite direction.
If a body emitting light is coming towards you, you will
find more waves in a certain direction than if it were
going from you; consequently, if you are approaching
a body emitting light, the waves will come at shorter
intervals, the vibration will be of shorter period, and the
light will be higher up in the spectrum—it will be more
blue.   If you are going away from the body, then the
rate is slower, the light is lower down on the spectrum,
and consequently more red.   By means of such varia-
tions in the positions of certain known lines, the actual
rate of approach of certain fixed stars to the earth has been
measured, and the rate of going away of certain other
fixed stars has also been measured.   Suppose we have
a gas which is glowing in a state of incandescence, all
the molecules are giving out light at a certain specified
rate of vibration; but some of these are coming towards
us at a rate much greater than seventeen miles a minute,
because the temperature is higher when the gas is glowing, and others are also going away at a much higher
rate than that.   The consequence is, that instead of

having one sharply defined line on the spectrum, instead of having light of exactly one bright colour, we have light which varies between certain limits. If the actual rate of the vibration of the molecules of the gas were marked down upon the spectrum, we should not get that single bright line there, but we should get a bright band overlapping it on each side. Lord Rayleigh calculated that, in the most favourable circumstances, the breadth of this band would not be less than one-hundredth of the distance between the sodium lines. It is precisely upon that experiment that the evidence of the exact similarity of molecules rests. We see, therefore, from the nature of the experiment, that we should get exactly the same results if the rates of vibration of all the molecules were not exactly equal, but varied within certain very small limits. If, for example, the rates of vibration varied in the same way as the heads of different men, then we should get very much what we get now from the experiment.

From the evidence of these two facts, then—the evidence that molecules are of the same weight and degree of vibration—all that we can conclude is that whatever differences there are in their weights, and whatever differences there are in their degrees of vibration, these differences are too small to be found out by our present modes of measurement. And that is precisely all that we can conclude in every similar question of science.

Now, how does this apply to the question whether it is possible for molecules to have been evolved by natural processes? I do not understand, myself, how, even supposing we knew that they were exactly alike

we could infer for certain that they had not been
evolved; because there is only one case of evolution
that we know anything at all about—and that we know
very little about yet—namely, the evolution of organized
beings. The processes by which that evolution takes
place are long, cumbrous, and wasteful processes of
natural selection and hereditary descent. They are processes which act slowly, which take a great lapse of ages
to produce their natural effects. But it seems to me quite
possible to conceive, in our entire ignorance of the
subject, that there may be other processes of evolution
which result in a definite number of forms,—those of the
chemical elements,—just as these processes of the evolution of organized beings have resulted in a greater
number of forms. All that we know of the ether shows
that its actions are of a rapidity very much exceeding
anything we know of the motions of visible matter. It
is a possible thing, for example, that mechanical conditions should exist according to which all bodies
must be made of regular solids, that molecules should all
have flat sides, and that these sides should all be of the
same shape. I suppose that it is just conceivable that
it might be impossible for a molecule to exist with two
of its faces different. In that case we know there would
be just five shapes for a molecule to exist in, and
these would be produced by a process of evolution.
The various forms of matter that chemists call
elements seem to be related one to another very
much in that sort of way; that is, as if they rose
out of mechanical conditions which only rendered it
possible for a certain definite number of forms to exist,
and which, whenever any molecule deviates slightly

from one of these forms, would immediately operate to set it right again. I do not know at all—we have nothing definite to go upon—what the shape of a molecule is, or what is the nature of the vibration it undergoes, or what its condition is compared with the ether ; and in our absolute ignorance it would be impossible to make any conception of the mode in which it grew up. When we know as much about the shape of a molecule as we do about the solar system, for example, we may be as sure of its mode of evolution as we are of the way in which the solar system came about ; but in our present ignorance all we have to do is to show that such experiments as we can make do not give us evidence that it is absolutely impossible for molecules of matter to have been evolved out of ether by natural processes.

The evidence which tells us that the molecules of a given substance are alike is only approximate. The theory leaves room for certain small deviations; and consequently if there are any conditions at work in the nature of the ether which render it impossible for other forms of matter than those we know of to exist, the great probability is that when by any process we contrive to sift molecules of one kind from molecules of another, these very conditions at once bring them back and restore to us a mass of gas consisting of molecules whose average type is a normal one.

Now I want to consider a speculation of an entirely different character. A remark was made about thirty years ago by Sir William Thomson upon the nature of certain problems in the conduction of heat. These problems had been solved by Fourier many years before in a beautiful treatise. The theory was that if you knew the degree of warmth of a body, then you could find what would happen to it afterwards; you would find how the body would gradually cool. Suppose you put the end of a poker in the fire and make it red hot, that end is very much hotter than the other end ; and if you take it out and let it cool, you will find that heat is travelling from the hot end to the cool end ; and the amount of this travelling, and the temperature at either end of the poker, can be calculated with great accuracy. This comes out of Fourier's theory. Now suppose you try to go backwards in time, and take the poker at any instant when it is about half cool, and say :

Does this equation give me the means of finding out what was happening before this time, in so far as the present state of things has been produced by cooling?' You will find the equation will give you an account of the state of the poker before the time when it came into your hands, with great accuracy up to a certain point ; but beyond that point it refuses to give you any more information, and it begins to talk nonsense. It is in the nature of a problem of the conduction of heat that it allows you to trace the forward history of it to any extent you like ; but it will not allow you to trace the history of it backward beyond a certain point. There is another case in which a similar thing happens. There is an experiment in that excellent manual, the 'Boy's Own Book,' which tells you that if you half fill a glass with beer, and put some paper on it, and then pour in water carefully, and draw the paper out without disturbing the two liquids, the water will rest on the beer. The problem then is to drink the

beer without drinking the water, and it is accomplished
by means of a straw. Let us suppose these two liquids
resting in contact; we shall find they begin to mix;
and it is possible to write down an equation exactly
of the same form as the equation for the conduction of heat, which would tell you how much water
had passed into the beer at any given time after the
mixture began. So that, given the water and the beer
half mixed, you could trace forward the process of mixing, and measure it with accuracy, and give a perfect
account of it; but if you attempt to trace that back you
will have a point where the equation will stop, and will
begin to talk nonsense. That is the point where you
took away the paper, and allowed the mixing to begin.
If we apply that same consideration to the case of the
poker, and try to trace back its history, you will find
that the point where the equation begins to talk nonsense is the point where you took it out of the fire.
The mathematical theory supposes that the process of
conduction of heat has gone on in a quiet manner,
according to certain defined laws, and that if at any
time there was a catastrophe, an event not included in the
laws of the conduction of heat, then the equation could
give you no account of it. There is another thing
which is of the same kind, namely the transmission of
fluid friction. If you take your tea in your cup, and
stir it round with a spoon, it will not go on circulating
round for ever, but will come to a stop; and the reason
is that there is a certain friction of the liquid against
the sides of the cup, and of the different parts of the
liquid with one another. The friction of the different
parts of a liquid or a gas is precisely a matter of

mixing. The particles which are going fast, and are in the middle, not having been stopped by the side, get mixed ; and the particles at the side going slow get mixed with the particles in the middle. This process of mixing can be calculated, and it leads to an equation of exactly the same sort as that which applies to the conduction of heat. We have, therefore, in these problems a natural process which consists in mixing things together, and this always has the property that you can go on mixing them for ever without coming to anything impossible ; but if you attempt to trace the history of the thing backward, you must always come to a state which could not have been produced by mixing, namely a state of complete separation.

Upon this remark of Sir William Thomson's, the true consequences of which you will find correctly stated in Mr. Balfour Stewart's book on the ' Conservation of Energy,' a most singular doctrine has been founded. These writers have been speaking of a particular problem on which they were employed at the moment. Sir William Thomson was speaking of the conduction of heat, and he said this heat problem leads you back to a state which could not have been produced by the conduction of heat. And so Professor Clerk Maxwell, speaking of the same problem, and also of the diffusion of gases, said there was evidence of a limit in past time to the existing order of things, when something else than mixing took place. But a most eminent man, who has done a great deal of service to mankind, Professor Stanley Jevons, in his very admirable book, the ' Principles of Science,' which is simply marvellous for the number of examples illustrating

logical principles which he has drawn from all kinds of
regions of science, and for the small number of mistakes
that occur in it, takes this remark of Sir W. Thomson's,
and takes out two very important words, and puts in
two other very important words He says : ' We have
here evidence of a limit of a state of things which could
not have been produced by the previous state of things
according to the known laws of nature.' It is not according to the known laws of nature, it is according to the
known laws of conduction of heat, that Sir William
Thomson is speaking ; and from this we may see the
fallacy of concluding that if we consider the case
of the whole universe we should be able, supposing we
had paper and ink enough, to write down an equation
which would enable us to make out the history of the
world forward, as far forward as we liked to go ; but if
we attempted to calculate the history of the world backward, we should come to a point where the equation
would begin to talk nonsense—we should come to a
state of things which could not have been produced
from any previous state of things by any known natural
laws. You will see at once that that is an entirely
different statement. The same doctrine has been used
by Mr. Murphy, in a very able book, the 'Scientific
Basis of Faith,' to build upon it an enormous super-
structure—I think the restoration of the Irish Church
was one of the results of it. But this doctrine is
founded, as I think, upon a pure misconception. It is
founded entirely upon forgetfulness of the condition
under which the remark was originally made. All
these physical writers, knowing what they were writing
about, simply drew such conclusions from the facts

which were before them as could be reasonably drawn. They say : 'Here is a state of things which could not have been produced by the circumstances we are at present investigating.' Then your speculator comes ; he reads a sentence, and says : 'Here is an opportunity for me to have my fling.' And he has his fling, and makes a purely baseless theory about the necessary origin of the present order of nature at some definite point of time which might be calculated. But, if we consider the matter, we shall see that this is not in any way a consequence of the theory of the conduction of heat ; because the conduction of heat is not the only process that goes on in the universe.

If we apply that theory to the case of the earth, we find that at present there is evidence of a certain distribution of temperature in the interior of it ; there is a certain rate at which the temperature increases as we go down ; and no doubt, if we made further investigations, we should find that if we went deeper an accurate law would be found, according to which the temperature increases in the interior.

Now, assuming this to be so, taking this as the basis of our problem, we might endeavour to find out what was the history of the earth in past times, and when it began cooling down. That is exactly what Sir William Thomson has done. When we attempt it, we find that there is a definite point to which we can go, and beyond which our equation talks nonsense. But we do not conclude that at that point the laws of nature began to be what they are ; we only conclude that the earth began to solidify. Now solidification is not a process of the conduction of heat, and so the thing cannot be given by our equation. That point is given definitely as a point of time, not with great accuracy, but still as near as we can expect to get it with such means of measuring as we have; and Sir William Thomson has calculated that the earth must have solidified at some time between a hundred millions and two hundred millions of years ago; and there we arrive at the beginning of the present state of things—the process of cooling the earth which is going on now. Before that it was cooling as a liquid, and in passing from the liquid to the solid state there was a catastrophe which introduced a new rate of cooling. So that by means of that law we do come to a time when the earth began to assume its present state. We do not find the time of the commencement of the universe, but simply of the present structure of the earth. If we went farther back, we might make more calculations and find how long the earth had been in a liquid state. We should come to another catastrophe, and say not that at that time the universe began to exist, but that the present earth passed from the gaseous to the liquid state. And if we went farther back still we should probably find the earth falling together out of a great ring of matter surrounding the sun and distributed over its orbit. The same thing is true of every body of matter: if we trace its history back, we come to a certain time at which a catastrophe took place; and if we were to trace back the history of all the bodies of the universe in that way, we should continually see them separating up into smaller parts. What they have actually done is to fall together and get solid. If we could reverse the process we should see them separating and getting fluid; and, as a limit to that, at an indefinite distance in past time, we should find that all these bodies would be resolved into molecules, and all these would be flying away from each other. There would be no limit to that process, and we could trace it as far back as ever we liked to trace it. So that on the assumption—a very large assumption—that the present constitution of the laws of geometry and mechanics has held good during the whole of past time, we should be led to the conclusion that at an inconceivably long time ago the universe did consist of ultimate molecules, all separate from one another, and approaching one another. Then they would meet together and form a great number of small, hot bodies. Then you would have the process of cooling going on in these bodies, exactly as we find it going on now. But you will observe that we have no evidence of such a catastrophe as implies a beginning of the laws of nature. We do not come to something of which we cannot make any further calculation; we find that however far we like to go back, we approximate to a certain state of things, but never actually get to it.

Here, then, we have a doctrine about the beginning of things. First, we have a probability, about as great as science can make it, of the beginning of the present state of things on the earth, and of the fitness of the earth for habitation; and then we have a probability about the beginning of the universe as a whole which is so small that it is better put in this form, that we do not know anything at all about it. The reason why I say that we do not know anything at all of the beginning of the universe is that we have no reason whatever for believing that the known laws of geometry and mechanics are exactly and absolutely true at present, or that they have been even approximately true for any period of time further than we have direct evidence of. The evidence we have of them is founded on experience; and we should have exactly the same experience of them now, if those laws were not exactly and absolutely true, but were only so nearly true that we could not observe the difference. So that in making the assumption that we may argue upon the absolute uniformity of nature, and suppose these laws to have remained exactly as they are, we are assuming something we know nothing about. My conclusion then is that we do know, with great probability, of the beginning of the habitability of the earth about one hundred or two hundred millions of years back, but that of a beginning of the universe we know nothing at all.

Now let us consider what we can find out about the end of things. The life which exists upon the earth is made by the sun's action, and it depends upon the sun for its continuance. We know that the sun is wearing out, that it is cooling; and although this heat which it loses day by day is made up in some measure, perhaps completely at present, by the contraction of its mass, yet that process cannot go on for ever. There is only a certain amount of energy in the present constitution of the sun; and when that has been used up, the sun cannot go on giving out any more heat. Supposing, therefore, the earth remains in her present orbit about the sun, seeing that the sun must be cooled down at some time, we shall all be frozen out. On the other hand, we have no reason to believe that the orbit of the earth about the sun is an absolutely stable thing. It has been maintained for a long time that there is a certain resisting medium which the planets have to move through; and it may be argued that in time all the planets must be gradually made to move in smaller orbits, and so to fall in towards the sun. But, on the other hand, the evidences upon which this assertion was based, the movement of Encke's comet and others, has been recently entirely overturned by Professor Tait. He supposes that these comets consist of bodies of meteors. Now it was proved a long time ago that a mass of small bodies travelling together in an orbit about a central body will always tend to fall in towards it, and that is the case with the rings of Saturn. So that, in fact, the movement of Encke's comet is entirely accounted for on the supposition that it is a swarm of meteors, without regarding the assumption of a resisting medium. On the other hand, it seems exceedingly natural to suppose that some matter in a very thin state is diffused about the planetary spaces. Then we have another consideration,—just as the sun and moon make tides upon the sea, so the planets make tides upon the sun. Consider the tide which the earth makes upon the sun. Instead of being a great wave lifting the mass of the sun up directly under the earth, it is carried forward by the sun's rotation; the result is that the earth, instead of being attracted to the sun's centre, is attracted to a point before the centre. The immediate tendency is to accelerate the earth's motion, and the final effect of this upon the planet is to make its orbit larger. That planet disturbing all the other planets, the consequence is that we have the earth gradually going away from the sun, instead of falling into it.[1]

In any case, all we know is that the sun is going out. If we fall into the sun then we shall be fried; if we go away from the sun, or the sun goes out, then we shall be frozen. So that, so far as the earth is concerned, we have no means of determining what will be the character of the end, but we know that one of these two things must take place in time. But in regard to the whole universe, if we were to travel forward as we have travelled backward in time, and consider things as falling together, we should come finally to a great central mass, all in one piece, which would send out waves of heat through a perfectly empty ether, and gradually cool itself down. As this mass got cool it would be deprived of all life and motion; it would be just a mere enormous frozen block in the middle of the ether. But that conclusion, which is like the one that we discussed about the beginning of the world, is one which we have no right whatever to rest upon. It depends upon the same assumption that the laws of geometry and mechanics are exactly and absolutely true; and that they will continue exactly and absolutely true for ever and ever. Such an assumption we have no right whatever to make. We may therefore, I think, conclude about the end of things that, so far as the earth is concerned, an end of life upon it is as probable as science can make anything; but that in regard to the universe we have no right to draw any conclusion at all.

[1] I learn from Sir W. Thomson that the ultimate effect of tidal deformation on a number of bodies is to reduce them to two, which move as if they were rigidly connected.

So far, we have considered simply the material existence of the earth; but of course our greatest interest lies not so much with the material life upon it, the organized beings, as with another fact which goes along with that, and which is an entirely different one—the fact of the consciousness that exists upon the earth. We find very good reason indeed to believe that this consciousness in the case of any organism is itself a very complex thing, and that it corresponds part for part to the action of the nervous system, and more particularly of the brain of that organized thing. There are some whom such evidence has led to the conclusion that the destruction which we have seen reason to think probable of all organized beings upon the earth will lead also to the final destruction of the consciousness that goes with them. Upon this point I know there is great difference of opinion amongst those who have a right to speak. But to those who do see the cogency of the evidences of modern physiology and modern psychology in this direction it is a very serious thing to consider that not only the earth itself and all that beautiful face of nature we see, but also the living things upon it, and all the consciousness of men, and the ideas of society, which have grown up upon the surface, must come to an end. We who hold that belief must just face the fact and make the best of it; and I think we are helped in this by the words of that Jew philosopher, who was himself a worthy crown to the splendid achievements of his race in the cause of progress during the Middle Ages, Benedict Spinoza. He said: 'The free man thinks of nothing so little as of death, and his wisdom is a meditation not of death but of life.' Our interest lies with so much of the past as may serve to guide our actions in the present, and to intensify our pious allegiance to the fathers who have gone before us and the brethren who are with us; and our interest lies with so much of the future as we may hope will be appreciably affected by our good actions now. Beyond that, as it seems to me, we do not know, and we ought not to care. Do I seem to say: 'Let us eat and drink, for to-morrow we die?' Far from it; on the contrary I say: 'Let us take hands and help, for this day we are alive together.'

The following note was afterwards published by the author ('Fortnightly Review,' vol. xvii. p. 793):—

The passage referred to from the 'Principles of Science' is as follows (vol. ii. p. 438):—

'For a certain negative value of the time the formulæ give impossible values, indicating that there was some initial distribution of heat which could not have resulted, according to known laws *of nature*, from any previous distribution.'

The words italicized are here inserted into a sentence from Tait's 'Thermo-dynamics,' p. 38. Had the words *conduction of heat* been used instead of *nature*, the sentence would have remained correct, but would not have led to the alarming inference that

'The theory of heat places us in the dilemma either of believing in creation at some assignable date in the past, or else of supposing that some inexplicable change in the working of natural laws then took place.'

It has been pointed out by Mr. Higgins that the ultimate effect of tides in the sun caused by the earth's attraction will be precisely similar to that of a resisting medium—that is, will diminish the orbit of the earth and increase its velocity; and that I was wrong in supposing the contrary effect. It results that the earth will certainly fall into the sun; but whether before or after the sun has cooled down so much as not to be able to support life on this planet, remains undetermined. The final conclusion remains therefore as before—that there must be an end, but whether by heat or by cold we cannot tell.

THE UNSEEN UNIVERSE. [1]

THE primary motive of this treatise is indicated by its second title: 'Physical Speculations on a Future State.' A sketch of the beliefs and yearnings of many different folk in regard to a life after death leads up to an attempt to find room for it within the limits of those physical doctrines of continuity and the conservation of energy which are regarded as the established truths of science. In this attempt it is necesssay to discuss the ultimate constitution of matter and its relation to the ether. When, by a singular inconsequence in writers possessing such power in their right minds of sound scientific reasoning, room has been found for a future life in the manner indicated above, it is discovered that there is room for a great deal more. Accordingly some of the main doctrines of the Christian religion are interpreted in relation to the authors' hypothesis, and placed in their appropriate niches. It will perhaps be convenient, therefore, if we consider these three things in their order: first, the desire for a future life; secondly, the physical speculations that make room for it; and lastly, that system, the seemingly innocent dried carcase of which is to be smuggled into our house at the same time, that it may peradventure finds means of resurrection.

[1] 'The Unseen Universe; or, Physical Speculations on a Future State.' London: Macmillan & Co. 1875. ['Fortnightly Review,' June, 1875.]

I.

It is often said that the universal longing for immortality among all kinds and conditions of men is a presumption that there is some future life in which this longing shall be satisfied. Let us endeavour, therefore, to find out in what this longing for immortality actually consists; whether the existence of it, when its nature is understood, can be explained on grounds which do not require it to have any objective fulfilment other than the life and the memory of those who come after us; and what relation it bears to the equally wide-spread dream or vision of a spiritual world peopled by supernatural or monstrous beings, ghosts and gods and goblins.

First, let us notice that all the words used to describ this immortality that is longed for are *negative* words: *im*-mortality, end-*less* life, *in*-finite existence. Endless life is an inconceivable thing, for an endless time would be necessary to form an idea of it Now it is only by a stretch of language that we can be said to desire that which is inconceivable. No doubt many persons say that they are smitten with an insatiable longing for the unattainable and ineffable; but this means that they feel generally dissatisfied and do not at all know what they want. Longing for deathlessness means simply *shrinking from death.* However or whenever we who live endeavour to realize an end to this healthy life of action in ourselves or in our brethren, the effort is a painful one; and the mind, in so far as it is healthy, tries to put it off and avoid it. The state of one who really wishes for death is firmly linked in our thoughts with the extreme of misery and wretchedness and disease;

and, in so far as it can be realized, we seem to feel that
such a one is fit to die.	In those cases of ripe old age
not hastened by disease, where the physical structure is
actually worn out, having finished its work right
honestly and well; where the love of life is worn out
also, and the grave appears as a bed of rest to the tired
limbs, and death as a mere quiet sleep from thought;
there also, in so far as we are able to realize the state
of the aged and to put ourselves in his place, death
seems to be normal and natural, a thing to be neither
sought nor shunned.	But such putting of ourselves in
the place of one to whom death is no evil must in all
cases be imperfect.	I cannot, in my present life and
motion, clearly conceive myself in so parlous a state
that no hope of better things should make me shrink
from the end of all.	However vividly I recall the feel-
ings of pain and weakness, it is the life and energy of my
present self that pictures them ; and this life and energy
cannot help raising at the same time combative instincts
of resistance to pain and weakness, whose very nature
it is to demand that the sun shall not go down upon
Gibeon until they have slain the Amalekites.	Nor can I
really and truly put myself in the place of the worn-
out old man whose consciousness may some day have a
memory of mine.	No force of imagination that I can
bring to bear will avail to cast out the youth of that
very imagination which endeavours to depict its latter
days ; no thoughts of final and supreme fatigue can
help suggesting refreshment and new rising after sleep.

If, then, we do not want to die now, nor next year,
nor the year after that, nor at any time that we can
clearly imagine ; what is this but to say that we want to

live for ever, in the only meaning of the words that we
can at all realize? It is not that there is any positive
attraction in the shadowy vistas of eternity, for the effort
to contemplate even any very long time is weariness
and vexation of spirit; it is that our present life, in so
far as it is healthy, rebels once for all against its own
final and complete destruction. And forasmuch as so
many and so mighty generations have in time past ended
in death their noble and brave battle with the elements,
that we also and our brethren can in nowise hope to
escape their fate, therefore we are sorely driven to find
some way by which at least the image of that ending
shall be avoided and set aside. As the fruit of this
search two methods have been found and practised
among men. By one method we detach ourselves from
the individual body and its actions which accompany
our consciousness, to identify ourselves with something
wider and greater that shall live when we as units shall
have done with living—that shall work on with new
hands when we, its worn-out limbs, have entered into
rest. The soldier who rushes on death does not know
it as extinction; in thought he lives and marches on with
the army, and leaves with it his corpse upon the battle-
field. The martyr cannot think of his own end because
he lives in the truth he has proclaimed; with it and
with mankind he grows into greatness through ever new
victories over falsehood and wrong. But there is another way. Since when men have died such orderly,
natural, and healthy activity as we have known in them
and valued their lives for has plainly ceased, we may
fashion another life for them, not orderly, not natural,
not healthy, but monstrous or *super*natural; whose

cloudy semblance shall be eked out with the dreams of
uneasy sleep or the crazes of a mind diseased. And it
is to this that the universal shrinking of men from
death, which is called a yearning for immortality, is
alleged to bear witness.

But whence now does it really come, and what is the
true lesson of it ?   Surely it is a necessary condition of
life that has desires at all that these desires should be
towards life and not away from it ; seeing how cheap
and easy a thing is destruction on all hands, and how
hard it is for race or unit to hold fast in the great
struggle for existence.   Surely our way is paved with
the bones of those who have loved life and movement
too little, and lost it before their time.   If we could
think of death without shrinking, it would only mean
that this world was no place for us, and that we should
make haste to be gone to leave room for our betters.
And therefore that love of action which would put
death out of sight is to be counted good, as a holy and
healthy thing (one word whose meanings have become
unduly severed), necessary to the life of men, serving to
knit them together and to advance them in the right.
Not only is it right and good thus to cover over and
dismiss the thought of our own personal end, to keep in
mind and heart always the good things that shall be done,
rather than ourselves who shall or shall not have the
doing of them ; but also to our friends and loved ones
we shall give the most worthy honour and tribute if we
never say nor remember that they are dead, but con-
trariwise that they have lived ; that hereby the
brotherly force and flow of their action and work may
be carried over the gulfs of death and made immortal

in the true and healthy life which they worthily had and used. It is only when the bloody hands of one who has fought against the light and the right are folded and powerless for further crime, that it is most kind and merciful to bury him and say, ‘The dog is dead.’

But for you, noble and great ones, who have loved and laboured yourselves not for yourselves but for the universal folk, in your time not for your time only but for the coming generations, for you there shall be life as broad and far-reaching as your love, for you life-giving action to the utmost reach of the great wave whose crest you sometimes were.

II.

Believing that every finite intelligence must be ‘conditioned in time and space,’ and therefore must have an ‘organ of memory’ and a ‘power of varied action,’ and consequently must be associated with a physical organism,—recognizing also that the world, as it is known at present, is made up of material molecules and of ether,—our authors frankly admit that no room is here to be found either for ghosts of the dead, or ‘superior intelligences,’ or bogies of any kind whatever. But modifying a hypothesis of Sir W. Thomson’s about the ultimate form of atoms and their relation to the ether, they find in a second ether the material where- with to refashion all these marvels which advancing knowledge had banished from the realm of reality. We may here, then, review with advantage for a short time the state of that border-land between the known and the unknown in physical science to which this ingenious hypothesis belongs; with the view of inquiring what measure of probability is to be attached to the modification of it which our authors propose.

Imagine a ring of india-rubber, made by joining together the ends of a cylindrical piece (like a lead pencil before it is cut), to be put upon a round stick which it will just fit with a little stretching. Let the stick be now pulled through the ring while the latter is kept in its place by being pulled the other way on the outside. The india-rubber has then what is called *vortex-motion*. Before the ends were joined together, while it was straight, it might have been made to turn round without changing position by rolling it between the hands. Just the same motion of rotation it has on the stick, only that the ends are now joined together. All the inside surface of the ring is going one way, namely the way the stick is pulled; and all the outside is going the other way. Such a vortex-ring is made by the smoker who purses his lips into a round hole and sends out a puff of smoke. The outside of the ring is kept back by the friction of his lips while the inside is going forwards; thus a rotation is set up all round the smoke-ring as it travels out into the air. If we half immerse a teaspoon in our tea and draw it across the surface, we may see two little eddies formed at the edges of the spoon. These eddies are really united by a sort of rope of fluid underneath the surface, which follows the shape of the spoon, and which has throughout the same motion of rotation that the india-rubber ring had when the stick was drawn through it; except that in this case only half a ring is formed, being cut off, as it were, by the surface of the liquid. In all these cases vortex-motion is produced by friction, and would be ultimately destroyed by friction. But, by way of an approximation to the study of water, men had been led to the conception of a *perfect liquid*; that is, a liquid absolutely free from friction, or (which is the same thing) offering no resistance to change of shape, or the sliding of one part over another. Water at rest behaves just as such a liquid would behave; but water in motion is altogether a different thing. Helmholtz found, by a wonderfully beautiful calculation, that in a perfect liquid where there is no friction it is impossible for vortex-motion to be generated or destroyed; in any part of the liquid where there is no vortex-motion no mechanical action can possibly start it; but where it once exists, there it is for ever, and no mechanical action can possibly stop it. A vortex-ring may move from place to place; but it carries with it the liquid of which it is composed, never leaving any particle behind, and never taking up any particle from the surrounding liquid. If we tried to cut it through with a knife, it would thin out like a stream of treacle, and the thinner it got the faster it would go round; so that if we multiplied together the number of revolutions in a second, and the number of square millimeters in the cross-section of the vortex-ring, we should always get the same product, not only in all parts of the ring, but through all time. Any portion of liquid which is rotating must form part of a vortex-ring, either returning into itself, after no matter how many knots and convolutions, or having its two ends cut off at the surface of the liquid. That such more complex forms of vortex-motion may exist is easily shown by making knots (to be left loose) in a piece of string, and then joining the ends : motion of rotation may be given to any part of it by rolling it between two fingers, and will be carried all over it. Such a knotted vortex-ring is figured on the cover of the ' Unseen Universe' for a fitting device.

Thus far Helmholtz, examining into the consequences of supposing that a fiction, serving to represent the actual properties of liquids at rest, holds good also in the case of motion. Here steps in Sir William Thomson with a brilliant conjecture. The ultimate atom of matter is required to be indestructible, to have a definite mass, and definite rates of vibration. A vortex-ring in a perfect liquid is indestructible, has a definite mass, and definite rates of vibration. Why should not the atom be a vortex-ring in a perfect liquid? If the whole of space were filled with an incompressible frictionless fluid in which vortex-rings once existed, at least some of the known phenomena of matter would be produced. Why should it nct be possible in this way to explain them all ?

The answer to this question is only to be got at by examining further into the consequences of the funda-mental supposition, until either the desired explanation of all phenomena is reached, or some clear discordance with observed results shows that the whole hypothesis is untenable. To this task, with splendid energy and insight, Sir William Thomson has applied himself; arriving at results which, if they are not the foundation of the final theory of matter, are at least imperishable stones in the tower of dynamical science.

Independently, however, of these results in the

theory of the motion of perfect liquids, and independently of the final success of the hypothesis itself, it has led to two very important ideas of physical explanation. First, there is the idea that matter differs from ether only in being another state or mode of motion of the same stuff; which suggests the hope that we may by-and-by get to know something about the method of evolution of atoms, and the reason why there are so many kinds of them and no more. It must not be supposed that in Sir W. Thomson's hypothesis the part of the ether is played simply by the universal frictionless fluid. Such a fluid, by the definition of it, offers no resistance to a change of shape of any part of it; but the actual ether which fills space is so elastic that the slightest possible distortion produced by the vibra tion of a single atom sends a shudder through it with inconceivable rapidity for billions and billions of miles. This shudder is Light. To account for such elasticity it has to be supposed that even where there are no material molecules the universal fluid is full of vortex-motion, but that the vortices are smaller and more closely packed than those of matter, forming altogether a more finely grained structure. So that the difference between matter and ether is reduced to a mere difference in the size and arrangement of the component vortex-rings. Now, whatever may turn out to be the ultimate nature of the ether and of molecules, we know that to some extent at least they obey the same dynamic laws, and that they act upon one another in accordance with these laws. Until, therefore, it is absolutely disproved, it must remain the simplest and most probable assumption that they are finally made of the same stuff—that the material molecule is some kind of knot or coagulation of ether.

Secondly, this hypothesis has accustomed us to the very important idea that the hardness, resistance, or elasticity of solid matter may be explained by the very rapid motion of something which is infinitely soft and yielding. This general view Sir William Thomson has illustrated by exceedingly beautiful experiments. One striking form is the complete enclosure of a gyroscope in a flat cylindrical box, with a sharp projecting edge, so that the motion of the contained wheel can only be perceived by the curious resistance to rotation of the box; which will balance itself on its edge on a piece of glass, and only tremble and stand firm when it is struck a violent blow with the hand. So also, if a chain hanging straight down be rapidly spun round, it becomes stiff and stark like a rigid rod. And, lastly, a solid suspended in the centre of a globe of water will, when the water is made to revolve rapidly, oscillate about its mean position as if it were fastened by a spring. All these things make one inclined to look to the rapid motion of something soft for explanation of hardness and stiffness; and the value of this explanation does not depend upon the ultimate success of the hypothesis of vortex-atoms.

But these things being admitted, it may perhaps not be too great a presumption in us to make some criticisms on the hypothesis itself. A true explanation describes the previous unknown in terms of the known; thus light is described as a vibration, and such properties of light as are also properties of vibrations are thereby explained. Now a perfect liquid is not a known thing, but a pure fiction. The imperfect liquids which approximate to it, and from which the conception is derived, consist of a vast number of small particles perpetually interfering with one another's motion. This molecular structure not only explains the fact that they behave like perfect liquids when at rest, but also makes it necessary that they should not behave like perfect liquids when in motion. Thus a liquid is not an ultimate conception, but is explained—it is known to be made up of molecules ; and the explanation requires that it should not be frictionless. The liquid of Sir William Thomson's hypothesis is continuous, infinitely divisible, not made of molecules at all, and it is absolutely frictionless. This is as much a mere mathematical fiction as the attracting and repelling points of Boscovitch.

The authors of the ' Unseen Universe ' modify the hypothesis in such a way as to dispose of this objection. They regard the atoms as not absolutely indestructible, but only very long-lived. Consequently it is not necessary for them that the universal liquid should be quite perfect, but only that its viscosity or friction should be exceedingly small—small enough to let the atoms keep going for billions of years when they are once started, with no appreciable change in their properties during the short time in which we can observe them. Thus, instead of a fiction, we have indeed a known thing, an imperfect liquid, by which to explain the molecules that are wanted to explain the properties of water. Can we, then, explain this universal imperfect liquid ? Certainly ; it consists of molecules inconceivably smaller than those of ordinary matter. But how to explain the molecules ? Why, clearly, they are vortex-rings in a liquid of still finer grain and less viscosity. Molecules, liquid, molecules, liquid, alternately for ever; each term of the infinite series being fully explained by the next following. Could anything be more satisfactory?

It is, moreover, to be observed that known facts about the ether and about atoms do lead us a very great way towards a conception of their relative structure. The experimental discoveries and the geometric insight of Faraday, and the application to these of mathematical analysis by Thomson, Helmholtz, and above all by Clerk Maxwell, have shown that the ether which was required for the theory of light is capable also of explaining magnetic and electric phenomena. Whatever that motion is which is periodically reversed in a ray of light, we have very strong evidence to show that the same motion is continuous along an electric current. This stream makes vortex-motion all round it, as if it were a stick drawn through india-rubber rings; and the vortex-rings are Faraday's 'lines of magnetic force.' The direction in which a small magnet will point indicates at any place the axis of rotation of the ether: thus, except in the neighbourhood of magnets or batteries, the ether in this country is all rotating in a plane rather tilted up on the north side. According to Maxwell's provisional conception, we may suppose that this rotation belongs to soft balls, all spinning the same way, and separated by smaller 'idle wheels,' which turn in the opposite direction. It is a continuous stream of these idle wheels that constitutes an electric current. Now there is great reason to believe that every material atom carries upon it a small electric current, if it does not wholly consist of this current. For, in the first place, every particle of a magnet is itself a magnet. Now, when a piece of iron is magnetized, there are two possible suppositions : either every particle is made into a magnet as it stands, having had no previous magnetism ; or else all the particles were originally magnets which neutralize one another because they were turned in all manner of directions, but which by the process of magnetizing have been made to approximate to the same direction. The latter supposition is conclusively picked out by experiment as the true one. Thus it seems that the molecule of iron is a magnet. If, however, the magnetism of the molecules were so much increased that they held each other tight, and so could not be turned round by ordinary magnetizing forces, it is shown that effects would be produced like those of diamagnetism. Faraday gave reasons for believing that all bodies are either ferromagnetic or diamagnetic. Next, the theory of Ampère, confirmed by many subsequent experiments and calculations, makes all magnetism to depend upon small electric currents. But magnetism is an affair of molecules ; if the molecules are groups of atoms, we find in this way good reason to suppose that all atoms carry upon them electric currents.

Three important sets of phenomena are (among many others) still unexplained—the action of molecules upon one another, the action of transparent bodies on light, and gravitation. The precise law of action of molecules on one another is in fact unknown, the inverse fifth power of the distance, proposed by Maxwell, having been given up on the evidence of later experiments.

The study of the mutual action of free small magnets in space offers mathematical difficulties which at present prevent us from saying whether a great number of these magnets would have such known properties of gases as depend upon the law of mutual action of molecules. Transparent bodies act upon light as if the ether in their interior were somewhat less elastic than the ether outside them. It is possible that this change of elasticity may be explained by the electric field surrounding their molecules, although the most powerful fields that we can produce have not yet been observed to have any such effect. There is something left for gravitation. In the theories of electric and magnetic action the motion of the 'idle wheels,' except in actual currents, is neglected in comparison with that of the revolving soft spheres. It is, perhaps, conceivable that in some way or other an explanation may be found in them for the relatively weaker force of gravitation. If—and what an if!—these three explanations were made out, we might reasonably suppose not merely that an atom *carries* an electric current, but that it *is* nothing else. We should thus be led to find an atom, not in the rotational motion of a vortex-ring, but in irrotational motion round a re-entering channel. It might well be that such motion, to be permanent, must have some definite relation to the size of the rotating spheres and their interstices, so that only certain kinds of atoms could survive. In this way we may get an explanation of the definite number of chemical elements, and of the fact that all the molecules of each are as near alike as we can judge.

The position is this. We know, with great probability, that wherever there is an atom there is a small electric current. Very many of the properties of atoms are explained by means of this current : we have vague hopes that all the rest will likewise be explained. If these hopes should be realized, we shall say that an atom is a small current. If not, we shall have to say that it is a small current and something else besides.

Of course, after all this, there is room for vortex-motion or other such hypotheses to explain the observed properties of the ether ; but in the last resort all these questions of physical speculation abut upon a meta-physical question. We are describing phenomena in terms of phenomena ; the objects we observe are groups of perceptions, and exist only in our minds ; the molecules and ether, in terms of which we describe them, are only still more complex mental images. Is there anything that is not in our minds of which these things are pictures or symbols ? and if so, what ?

Our authors reply that matter and energy possess this external reality, because they cannot be created or destroyed by us ; the quantity of each is fixed and invariable. The argument is better than most that belong to this question, but it will not hold water for a moment. Every quantitative relation among pheno-mena can be put into a form which asserts the con-stancy of some quantity which can be calculated from the phenomena. ' Gravitation is inversely as the square of the distance for the same two bodies ; ' this may be also said in the form, ' gravitation multiplied by the square of the distance is constant for the same two bodies.' ' Pressure varies as density, in a perfect gas at the same temperature,' may be also expressed, ' pressure divided by density is constant in a perfect

gas at the same temperature.' But this does not make
the quotient of pressure by density to be an external
reality transcending phenomena. It is entirely beside
the question, as we may see in another way. A dream
is a succession of phenomena having no external reality
to correspond to them. Do we never dream of things
that we cannot destroy?

So the fact that matter, as a phenomenon, is not to
be increased or diminished in quantity, has nothing to
say to the question about the existence of something
which is not matter, not phenomenon at all, but of which
matter is the symbol or representative. The answer
to this question is only to be found in the theory of
sensation; which tells us not merely that there is a
non-phenomenal counterpart of the material or pheno-
menal world, but also in some measure what it is made
of. Namely, the reality corresponding to our perception of the motion of matter is an element of the complex
thing we call feeling. What we might perceive as a
plexus of nerve-disturbances is really in itself a feeling;
and the succession of feelings which constitutes a man's
consciousness is the reality which produces in our minds
the perception of the motions of his brain. These elements of feeling have relations of *nextness* or contiguity
in space, which are exemplified by the sight-perceptions
of contiguous points; and relations of succession in
time which are exemplified by all perceptions. Out of
these two relations the future theorist has to build up
the world as best he may. Two things may, perhaps,
help him. There are many lines of mathematical
thought which indicate that distance or quantity may
come to be expressed in terms of *position* in the wide

sense of the *analysis situs*. And the theory of space-curvature hints at a possibility of describing matter and motion in terms of extension only.

So much for the vortex-atom, its relation to the present state of science, and the prospects of physical speculation. We propose now to follow our authors further; to examine their hypothesis of a second ether, and to see what good it can do them.

There are four ways of accounting for the too small number of stars of low magnitudes without assuming that light is absorbed by the ether. In the first place, the calculation assumes that stars are distributed with approximate uniformity over infinite space. So far is this from being true, that we know the vast majority of stars that we can see to belong to a single system, of which the nebulæ also are members, and which occupies a finite portion of space. It is very probable that around and beyond this, to distances vaster even than its vast dimensions, there are regions nearly devoid of stars. If other such systems do anywhere exist, they may well be too far off to be seen at all. The method of Struve has, indeed, been beautifully applied by Mr. Charles S. Peirce to the richer materials now at hand with the view of determining approximately the shape of the solar galaxy and the mode of distribution of stars in it. Secondly, a great amount of light must be stopped by the dark bodies of burnt-out suns. Thirdly, space contains gaseous matter in a state of extreme diffusion—not too rare, however, to produce an effect in distances so enormous as we have here to consider. Lastly, the possible curvature and finite extent of space have been suggested by Zöllner as an escape from the reasoning of Olbers and Struve. Of these four the first is undoubtedly the true account of the matter, and will supply us with trustworthy knowledge of the contents of surrounding space.

But if the ether did absorb light, what would this mean? Vibratory motion of solids, which is really a molecular disturbance, is absorbed by being transformed into other kinds of molecular motion, and so may finally be transferred to the ether. There is no reason why vibratory motion of the ether should not be transformed into other kinds of ethereal motion; in fact, there is no reason why it should not go to the making of atoms. Of course there is equally no reason why it should; but we present this speculation to anybody who wants the universe to go on for ever.

Apart from this, however, the laws of motion and the conservation of energy are very general propositions which are as nearly true as we can make out for gross bodies, and which, being tentatively applied to certain motions of molecules and the ether, are found to fit. There is nothing to tell us that they are absolutely exact in any particular case, or that they are everywhere and always true. If it were shown conclusively that energy was lost from the ether, it would not at all follow that it was handed on to anything else. The right statement might be that the conservation of energy was only a very near approximation to the facts.

It is perhaps hardly necessary to say that the experiment of Tait and Balfour Stewart, who found that a disc was heated by rapid rotation *in vacuo*, though of the first importance in itself, by no means bears upon the question of the internal friction of the ether. That a

molecule in travelling through the ether should be made
to vibrate is just what we might expect; the only
wonder is that it gets through with so little resistance.
But this is a transfer of energy of translation of a molecule into energy of vibration; a task to which *one* ether
is entirely competent.

Far greater, indeed, is the work which the second
ether has to perform : nothing less than the fashioning
of a 'spiritual body.' While our consciousness proceeds
pari passu with molecular disturbance in our brains,
this molecular disturbance agitates the first ether, which
transfers a part of its energy to the second. Thus is
gradually elaborated an organism in that second or
unseen universe, with whose motions our consciousness
is as much connected as it is with our material bodies.
When the marvellous structure of the brain decays, and
it can no more receive or send messages, then the
spiritual body is replete with energy, and starts off
through the unseen, taking consciousness with it, but
leaving its molecules behind. Having grown with the
growth of our mortal frame, and preserving in its structure a record of all that has befallen us, it becomes an
organ of memory, linking the future with the past, and
securing a personal immortality.

Can another body, then, avail to stay the hand of
death, and shall man by a second nervous system escape
scot free from the ruin of the first? We think not.
The laws connecting consciousness with changes in the
brain are very definite and precise, and their necessary
consequences are not to be evaded by any such means.
Consciousness is a complex thing made up of elements,
a stream of feelings. The action of the brain is also a

complex thing made up of elements, a stream of nerve-messages. For every feeling in consciousness there is at the same time a nerve-message in the brain. This correspondence of feeling to nerve-message does not depend on the feeling being part of a consciousness, and the nerve-message part of the action of a brain. How do we know this? Because the nervous system of animals grows more and more simple as we go down the scale, and yet there is no break that we can point to and say, ' above this there is consciousness or something like it; below there is nothing like it.' Even to those nerve-messages which do not form part of the continuous action of our brains, there must be simultaneous feelings which do not form part of our consciousness. Here, then, is a law which is true throughout the animal kingdom; nerve-message exists at the same time with feeling. Consciousness is not a simple thing, but a complex; it is the combination of feelings into a stream. It exists at the same time with the combination of nerve-messages into a stream. If individual feeling always goes with individual nerve-message, if combination or stream of feelings always goes with stream of nerve-messages, does it not follow that when the stream of nerve-messages is broken up, the stream of feelings will be broken up also, will no longer form a consciousness? does it not follow that when the messages themselves are broken up, the individual feelings will be resolved into still simpler elements? The force of this evidence is not to be weakened by any number of spiritual bodies. Inexorable facts connect our consciousness with this body that we know; and that not merely as a whole, but the parts of it are connected severally with parts of our brain-action. If there is any similar connexion with a spiritual body, it only follows that the spiritual body must die at the same time with the natural one.

Consider a mountain rill. It runs down in the sunshine, and its water evaporates; yet it is fed by thousands of tiny tributaries, and the stream flows on. The water may be changed again and again, yet still there is the same stream. It widens over plains, or is prisoned and fouled by towns; always the same stream; but at last

> 'even the weariest river
> Winds somewhere safe to sea.'

When that happens, no drop of the water is lost, but the stream is dead. ?

III.

Our authors 'assume, as absolutely self-evident, the existence of a Deity who is the Creator of all things.' They must both have had enough to do with examinations to be aware that 'it is evident' means 'I do not know how to prove.' The creation, however, was not necessarily a direct process; the great likeness of atoms gives them the 'stamp of the manufactured article,' and so they must have been made by intelligent agency, but this may have been the agency of finite and conditioned beings. As such beings would have bodies made of one or other of the ethers, this form of the argument escapes at least one difficulty of the more common form, which may be stated as follows:—'Because atoms are exactly alike and apparently indestructible, they must at one time have come into existence out of nothing. This can only have been effected by the agency of a conscious mind not associated with a material organism.' Forasmuch as the momentous character of the issue is apt to blind us to the logic of such arguments as these, it may not be useless to offer for consideration the following parody. 'Because the sea is salt and will put out a fire, there must at one time have been a large fire lighted at the bottom of it. This can only have been effected by the agency of the whale who lives in the middle of Sahara.' But let us return to our finite intelligences having ethereal bodies, who made the atomic vortex-rings out of ether. With such a machinery it seems a needless simplification to adopt Prout's hypothesis, and suppose that the sixty-three elements are compounded of one simpler form of matter. Rather let us contemplate the reposeful picture of the universal divan, where these intelligent beings whiled away the tedium of eternity by blowing smoke-rings from sixty-three different kinds of mouths. We may suppose, if we like, that the intelligent beings were all alike, and each had sixty-three mouths; or that each was so constituted in his physical or moral nature that he could or would pull only sixty-three faces. How lofty must have been the existence of such a maker and master of grimace! How fertile of resource is the theologic method, when it once has clay for its wheel!

As the permanence of matter proves the existence of an external reality, a substance in which all things consist, so the conservation of energy points to a principle of motion, coming out of the unconditioned, entering into the visible universe and obeying its laws, to pass back finally into the unseen world. But, further, the fact that organisms large enough to be visible have not yet under the conditions of the laboratory been produced from inorganic matter, shows that life is a great mystery, penetrating into the depths of the arcana of the universe, proceeding from substance and energy and yet not identical with either. The reader will see what this points to. It is clear that the good old gods of our race—sun, sky, thunder, and beauty—are to be replaced by philosophic abstractions—substance, energy, and·life, under the patronage respectively of the persons of the Christian Trinity. But why are we to stay here? Is not neurility, the universal function of nerves, as much a special and distinct form of life as life is a distinct form of energy? And over against these physical principles, absolutely separate and distinct from them, stands Consciousness, which cannot be left out of a fair estimate of the world. It would seem fitting that the presidency and patronage of the nerves should be assigned to the modern Isis as her portion. While if, as Von Hartmann says, Consciousness is the great mistake of the universe, it will not unsuitably fall to the care of the devil. In this way we shall save the odd number (*numero deus impare gaudet*), and give a certain historical completeness to our representation.

But why does a material so plastic present itself in this identical shape? Why this particular trinity of the great Ptah, Horus the Son, and Kneph the Wind-god, retained and refurbished by bishops of Alexandria and Carthage out of the wrecks of Egyptian superstition? Not because it is contained in the unseen universe, but because we were born in a particular place. If you, however, choose to find one thing in the chain of ethers, we may quite lawfully find another. If there is room in the unseen universe for the harmless pantheistic deities which our authors have put there, room may also be found for the goddess Kali, with her obscene rites and human sacrifices, or for any intermediate between these. Here is the clay : make your images to your heart's desire !

When Mohammed was conquering Arabia, a certain tribe offered to submit if they should be spared the tribute and service in the holy war, and if they might keep their idol Lat for a year. The prophet agreed, and began to dictate to his scribe the terms of the treaty. When it came to the permission of idolatry he paused and looked on the ground. The envoys were impatient, and repeated the article. Then arose Omar, and turned upon them furious. 'You have soiled the heart of the Prophet,' he said ; 'may God fill your hearts with fire !' 'I refuse the treaty,' said Mohammed, looking up. 'Let us keep Lat only six months, then,' pleaded the envoys. 'Not another hour,' said the Prophet ; and he drove them out and subdued them.

'Only for another half-century let us keep our hells and heavens and gods.' It is a piteous plea ; and it has soiled the heart of these prophets, great ones and blessed, giving light to their generation, and dear in particular to our mind and heart. These sickly dreams of hysterical women and half-starved men, what have they to do with the sturdy strength of a wide-eyed hero who fears no foe with pen or club ? This sleepless vengeance of fire upon them that have not seen and have not believed, what has it to do with the gentle patience of the investigator that shines through every

page of this book, that will ask only consideration and
not belief for anything that has not with infinite pains
been solidly established?[1] That which you keep in
your hearts, my brothers, is the slender remnant of a
system which has made its red mark on history, and
still lives to threaten mankind. The grotesque forms
of its intellectual belief have survived the discredit of
its moral teaching. Of this what the kings could bear
with, the nations have cut down; and what the nations
left, the right heart of man by man revolts against day
by day. You have stretched out your hands to save
the dregs of the sifted sediment of a residuum. Take
heed lest you have given soil and shelter to the seed of
that awful plague which has destroyed two civilizations
and but barely failed to slay such promise of good as
is now struggling to live among men.

[1] [Some time after this Essay was first published a reviewer more studious
of effect than accuracy said of Prof. Clifford : 'He invokes "the sleepless
vengeance of fire" upon those who do not share his unbelief!' The grotesque-
ness of the misrepresentation makes it worth while to record it, but not in
the language which strictly would be fitting to mark its recklessness or
impudence.]

THE PHILOSOPHY OF THE PURE SCIENCES.[1]

I.—STATEMENT OF THE QUESTION.

ON entering this room and looking rapidly round, what do I see ? I see a theatre, with a gallery, and with an arrangement of seats in tiers. I see people sitting upon these seats, people with heads more or less round, with bodies of a certain shape ; sitting in various positions. Above I see a roof with a skylight, and a round disc evidently capable of vertical motion. Below I see the solid floor supporting us all. In front of me I see a table, and my hands resting upon it. In the midst of all these things I see a void space, which I can walk about in if I like. The different things I have mentioned I see at various distances from one another, and from me ; and (now that the door is shut) I see that they completely enclose this void space, and hedge it in. My view is not made of patches here and there, but is a continuous boundary going all round the void space I have mentioned. All this I see to exist at the same time ; but some of you are not sitting quite still, and I see you move ; that is to say, I see you pass from one position into another by going through an infinite series of intermediate positions. Moreover, when I put my hands on the table, I feel a hard flat horizontal surface at rest, covered with cloth.

Have I spoken correctly in making these assertions ?

[1] Lectures delivered at the Royal Institution in March, 1873.

Yes, you will say, this is on the whole just what I ought to have seen and felt under the circumstances. With the exception of one or two points expressed in too technical a form, this is just the sort of language that a witness might use in describing any ordinary event, without invalidating his testimony. You would not say at once, 'This is absurd; the man must not be listened to any longer.' And if, having been precisely in my situation, you wished to describe facts with the view of drawing inferences from them—even important inferences—you would make all these statements as matter of your own direct personal experience; and if need were, you would even testify to them in a court of law.

And yet I think we shall find on a little reflection that not one of these statements can by any possibility have been strictly true.

'I see a theatre.' I do not; the utmost I can possibly see is two distinct curved pictures of a theatre. Upon the two retinas of my eyes there are made pictures of the scene before me, exactly as pictures are made upon the ground glass in a photographer's camera. The sensation of sight which I get comes to me at any rate through those two pictures; and it cannot tell me any more, or contain in itself any more, than is in those two pictures. Now the pictures are not solid; each of them is simply a curved surface variously illuminated at various parts. Whereas, therefore, I think I see a solid scene, having depth, and relief, and distance in it, reflection tells me that I see nothing of the kind; but only (at the most) two distinct surfaces, having no depth and no relief, and only a kind of distance which is quite different from that of the solid figures before me. You will say, probably, that this is only a quibble on two senses of the word ' see.' Whether it is so or not makes no difference to our subsequent argument; and yet I think you will admit that the latter sense, in which I do not see the solid things, is the more correct one. For the question is not about what is there, but about what I see. Now exactly the same sensation can be produced in me by two slightly different pictures placed in a stereoscope—I say exactly the same; because if I had sufficiently accurate coloured photographs of this room properly illuminated, the rays of light converging on every part of each of my retinas might be made exactly the same as they are now; and the sensation would therefore not only appear to be the same but would actually be the same. I should think I saw a solid scene; and I should not be seeing one. Now to see, and to see what is actually there, are two different things.

Again, ' I see people with heads more or less round.' —I cannot see your heads; I can only see your faces. I must have imagined the rest. But just consider what it is that I have imagined. Is it merely that besides what I do see I have added something that I might see by going round to the other side? No, there is more than that. The complete sensation which I have of a human head when I look at one is not merely something which I do not see now, but something which I never could see by any possibility. I have the sensation of a solid object, and not of a series of pictures of a solid object. Although that sensation may be really constructed out of a countless number of possible pictures, yet it is not like any of them. I imagine to myself, and seem to see the other side of things, not as it would look if viewed from beyond them, but as it would look if viewed from here. I seem to see the back of your head, not as it would look if I got behind you, but as if I saw it through your face from the spot where I am standing; and that, you know, is impossible.

I seem to see all these objects as existing together. But really as a matter of fact I move my eyes about and see a succession of small pictures very rapidly changed. Each of my eyes has six muscles which pull it about, and if I knew which of these muscles were moving, and how fast, at any moment, I should get information about the direction in which my eye was looking at the time. Now it is only a very small part of the scene before me that I can really see distinctly at once ; so that I have really seen a panorama, and not the one large picture that I imagined ; and yet while looking at the small portion which I can really see distinctly, I think I see distinctly the whole room.

Again, I seem to see that in some directions, at least, this void space in the middle is completely bounded —the surface of the floor, for example, which bounds it, appears to be completely filled up and continuous, to have no breaks in it. And when you move I seem to see you go *continuously* from one position to another through an infinite series of intermediate positions. Now, quite apart from the question whether these conclusions are true or not, it can be made out distinctly that I could not possibly see either the surface of a thing, or a motion, as continuous ; for the sensitive portion of my retina, which receives impressions, is not

itself a continuous surface, but consists of an enormously large but still finite number of nerve filaments
distributed in a sort of network. And the messages
that go along my nerves do not consist in any continuous action, but in a series of distinct waves succeeding
one another at very small but still finite intervals. All
I can possibly have seen therefore at any moment is a
picture made of a very large number of very small
patches, exceedingly near to one another, but not actually touching. And all I can have seen as time passed
is a succession of such distinct pictures coming rapidly
after one another. You know that precisely as the
stereoscope is made to imitate the property of my two
eyes out of which I imagine solid things, so another
instrument has been constructed to imitate that property of my nerves out of which I imagine continuous
motion. The instrument is called the Zoetrope, or
Wheel of Life. It presents to you a succession of distinct pictures coming after one another at small intervals; and the impression produced by that series is
precisely the impression of one thing in continuous
motion.

Let us now put shortly together what we have said
about this sensation of sight. I shall use the word
mosaic to represent a few disconnected patches which
a painter might put down with a view of remembering
a scene he had no time to sketch. Then, I seem to see
a large collection of solid objects in continuous motion.
The utmost I can really see is a panorama painted in
mosaic and shown in a wheel of life. I do not know
that my direct perception amounts to so much; but it
cannot possibly amount to more. What it really does

amount to must be reserved for subsequent discussion. At any rate I must have imagined the rest.

Lastly, when I put my hands on the table, I feel a hard, flat, horizontal surface at rest, covered with cloth. Now there are three things that really happen. First, there is a definite kind of irritation of certain organs of my skin, called papillæ. It is that irritation that makes me say *cloth*. Secondly, certain of my muscles are in a state of compression, and they tell me that. Thirdly, I make a certain muscular effort which is not followed by motion. This is all that I can really feel; but those three things do not constitute a hard, flat, horizontal surface covered with cloth. As before, I must have imagined the rest.

Do not suppose that I am advocating any change in our common language about sensation. I do not want anybody to say, for instance, instead of, 'I saw you yesterday on the other side of the street,' ' I saw a series of panoramic pictures in a sort of mosaic, of such a nature that the imaginations I constructed out of them were not wholly unlike the imaginations I have constructed out of similar series of panoramic pictures seen by me on previous occasions when you were present.' This would be clumsy, and it would not be sufficient. And yet I cannot help thinking that in certain assemblies, when some of those who are present are in an exalted state of emotional expectation, and the lights are low, even this roundabout way of putting things might be, to say the least, a salutary exercise.

But the conclusion I want you to draw from all this that we have been saying is that there are really two distinct parts in every sensation that we get. There is a message that comes to us somehow; but this message is not all that we apparently see and hear and feel. In every sensation there is, besides the actual message, something that we imagine and add to the message. This is sometimes expressed by saying that there is a part which comes from the external world and a part which is supplied by the mind. But however we express it, the fact to be remembered is that not the whole of a sensation is immediate experience (where by immediate experience I mean the actual message—whatever it is—that comes to us); but that this experience is supplemented by something else which is not in it. And thus you may see that it is a perfectly real question, 'Where does this supplement come from?' This question has been before philosophers for a very long time; and it is this question that we have to discuss.

But first of all we must inquire a little further into the nature of the supplement by which we fill in our experience. When I fill in my experience of this room in the way that I have described, I do not do so at random, but according to certain rules. And in fact I generally fill it in *right*; that is to say, from the imaginations that I have built up I can deduce by rules certain other experiences which would follow from actions of a definite sort. When I seem to see a solid floor, I conclude that if I went there I could feel it as I do the table. And upon trial these conclusions in general turn out right. I cannot therefore have filled in my experience at random, but according to certain rules. Let us now consider what are a few of these rules.

In the first place, out of pictures I have imagined

solid things.   Out of space of two dimensions, as we call
it, I have made space of three dimensions, and I imagine
these solid things as existing in it; that is to say, as
having certain relations of distance to one another.
Now these relations of distance are always so filled in as
to fulfil a code of rules, some called common notions,
and some called definitions, and some called postulates,
and some assumed without warning, but all somehow
contained in Euclid's 'Elements of Geometry.' For
example, I sometimes imagine that I see two lines in a
position which I call parallel.   Parallelism is impossible
on the curved pictures of my retina ; so this is part of
the filling in.   Now whenever I imagine that I see a
quadrilateral figure whose opposite sides are parallel, I
always fill them in so that the opposite sides are also
equal.   This equality is also a part of the filling in,
and relates to possible perceptions other than the one
immediately present.   From this example, then, you can
see that the fundamental axioms and definitions of
geometry are really certain rules according to which we
supplement or fill in our experience.

Now here is a rather more complicated example.   If
I see a train going along and a man moving inside of it,
I fill in the motion of the train as continuous out of a
series of distinct pictures of it ; and so also I fill in the
motion of the man relatively to the train as continuous.
I imagine all motions, therefore, according to the rule
of continuity ; that is, between the distinct pictures
which I see, I insert an infinite number of intermediate
pictures.   Moreover, both of these motions are imagined
in accordance with the laws of geometry ; that is to say,
they are imagined so that the relations of distance at any

instant obey those laws. But now I may, if I like, consider, besides the motion of the train and the motion of the man relative to it, the motion of the man relative to me, as if there were no train ; and this like the other motions is part of the filling in. But I always fill this in in such a way that the three motions—of the train by itself, of the man by himself, and of the man relatively to the train—satisfy certain rules, by which one can be found when the other two are given. These rules are called the laws of kinematic, or of the pure science of motion.

Then we may say, to begin with, that we supplement our experience in accordance with certain rules ; and that some of these rules are the foundations of the pure sciences of Space and Motion.

Instead of Space and Motion, many people would like to say Space and Time. But in regard to the special matter that we are considering, it seems to me, for reasons which I do not wish to give at present, to be more correct to say that we imagine time by putting together space and motion, than that we imagine motion by putting together space and time.

There are other rules, besides those of space and motion, according to which we fill in our experience. One of these rules I may call the continuity of things. I can see this table, and feel it, and hear a sound when I strike it. The table is an imagination by which I fill in a great variety of different experiences. It is what I call a thing. Now, if I come into this room again, and have any experience of the table, I shall fill it in in such a way as to imply that the same variety of experiences might be combined again ; that is, I shall imagine the thing to be persistent. But this rule will not apply universally, and I do not always observe it. Because I have seen a tree without leaves in the winter, I do not in the summer fill in my experience of the trunk with imagination of leafless branches above. But I do fill in the two experiences with an imagination of an infinite series of gradual intermediate changes. Some people divide this rule into two—the persistence of substance and the continuity of qualities. I prefer to make one rule, and to call it the continuity of things. Things— that is to say, combinations of possible experience—are not persistent, but they change continuously in the imagination by which we fill up that experience. Or we may say that experience at any one time is always so filled in as to aggregate together the possible perceptions implied by the result into groups which we call things ; and that experience of a period of time is always so filled in that things change only in a continuous manner.

Another rule of the supplement which we imagine is that which provides that these changes of things shall take place according to a certain uniformity. The simplest case of this is when the same experience is repeated, and we fill up the changes subsequent to the second experience so that they. shall be the same as those subsequent to the first. It is not necessary that the experience should be actually repeated ; it may only be filled up in the same way. The uniformity, however, which is involved in this law is a much more compli- cated thing than this simple case. I can only say here that experience is filled up always so as to make the imagined history of things exhibit *some* uniformity ; but the definiteness of this varies in different individuals and at different times. Some people prefer to call this the law of causation, and to say that we always supplement our experiences in such a way that every event has a cause or causes which determine it, and effects which flow from it.

Now all this filling up that we have been considering happens directly in the sensations that I get from day to day, just as I get them. (It is convenient to use the word sensation, as meaning the *whole* phenomenon, not only the immediate experience, but also the supplement.) But if I want to talk to you about them, or if, advancing upon that practice, I talk to myself about them, then I am obliged to use *language*, or to represent them by signs ; and this requires me to group them in a new manner. I have to make imaginations not of things, but of whole series of things, of relations of these to one another, and combinations of the relations. I have to construct, in fact, what I shall call for shortness the apparatus of thought—the means by which I talk to myself. For there seems reason to think that the conceptions which correspond to general terms—names of a class, or of an abstract relation—are first rendered necessary by the language which expresses them.[1] But however that may be, this new world of conceptions is not made wholly at random, but satisfies certain laws. For example, in order to describe a certain group of things, I introduce the very complicated conception *six*, and say there are six of them. Now, whenever this is done in the case of two groups, giving rise to the conceptions *six* and *three*, it is possible to apply the same

---

[1] See this view ably defended in Professor Max Müller's Lectures, delivered at the Royal Institution in April, 1873, and since published in *Fraser's Magazine.*

process to the group compounded of those two, and it always gives rise to the conception *nine*.   Here, then, is a law of combination to which the world of conceptions has to conform.   And another is this: If every individual which belongs to the class A belongs also to the class B, and if every individual which belongs to the class B belongs also to the class C, then always every individual which belongs to the class A belongs also to the class C. Rules like these which regulate the world of conceptions, built out of our sensations, are also said to belong to the pure sciences; and the two examples which I have chosen belong respectively to the sciences of Number and Logic.

There may be other kinds of rules according to which experience is supplemented and sensations are built up into conceptions; but I am not aware of any more kinds, and perhaps those that I have mentioned will be sufficient for our purpose.   I will just state again the names of the sciences which consist in these three groups:—

The rules about Space and Motion constitute the pure sciences of Geometry and Kinematic.

The rules about Things and Uniformity have been said to belong to a pure science of Nature.

The rules about Numbers and Classes constitute the pure sciences of Arithmetic and Formal Logic.

But for the present let us confine our attention to the first group of rules, those which relate to space and motion.   There is one other property of them which we have to consider, besides the fact that our experience is filled up in accordance with them.   I have already mentioned this property, but only in passing.

It is that in general this filling in of experience is *right*: and that, so far as these rules are concerned, it is not only right in general, but always right. That is to say, if from the sensation which is made by the filled-up experience we predict certain other perceptions as consequent upon our actions, these predictions will actually be fulfilled. To take the example we considered before, I always imagine a parallelogram so that its opposite sides are equal. Now the conclusion from this is that if I go to the parallelogram and apply one of the sides to the other, I shall not perceive any difference. The rule by which I supplement my perception is also a true statement about objects ; it is capable of a certain kind of verification, and it always stands this test.

Here, however, I could use the word *equal* only in its practical sense, in which two things are equal when I cannot perceive their difference ; not in its theoretical sense, in which two things are equal when they have no difference at all. But there has been for ages a conviction in the minds of men that these rules about space are true objectively in the exact or theoretical sense, and under all possible circumstances. If two straight lines are drawn perpendicular to the same plane, geometers would have told you for more than two thousand years that these straight lines may be prolonged for ever and ever without getting the least bit nearer to one another or further away from one another ; and that they were perfectly certain of this. They knew for certain that the sum of the angles of a triangle, no matter how big or how small it was, or where it was situated, must always be exactly equal to two right angles, neither more nor less. And those who were philosophers as well as geometers knew more than this. They knew not only that the thing was true, but that it could not possibly have been otherwise; that it was necessarily true. And this means, apparently, not merely that I know that it must be, but that I know that you must know that it must be.

The case of arithmetical propositions is perhaps more easily comprehended in this respect. Everybody knows that six things and three things make nine things at all possible times and places; you cannot help seeing not only that they do always without exception make nine things, but that they must do so; and that the world could not have been constructed otherwise. For to those ingenious speculations which suppose that in some other planet there may always be a tenth thing inevitably suggested upon the union of the six and the three, so that they cannot be added together without making ten; to these, I say, it may be replied that the words *number* and *thing*, if used at all, must have different meanings in that planet. The reply is important, and I shall return to it in a subsequent lecture.

Locke and Hume gave explanations of the existence of two of these general rules which I have put into my second group. Locke explained the notion of substance, the notion that a *thing* means something more than an aggregate of possible perceptions, by the fact that we are accustomed to get these perceptions all together; by this *custom* they are welded or linked together, and our imagination of the thing is then this connected structure of perceptions, which is called up as a whole whenever one or more of the component perceptions is called up. Having thus by custom formed

the complete sensation which we have of the thing, we
suppose that this is a message, like the actual perceptions, and comes from something outside.   That something is the substance.   Locke did not admit that this
supposition is right, and that the linking together of
messages is really itself a message; but still he thought
there was something outside to correspond to this
linking.   Hume explained in the same way the rule of
causation.   He said we get it from being accustomed to
perceive one event following another ; so that these two
perceptions got linked together, and when one of them
occur alone, we fills it in with the other one.   And
then, regarding this link, produced only by custom, as
if it were a message from somewhere, like the simple
perceptions, we give it the name of causation.

These explanations agree in saying that the supplement of experience is made up of past experience, together with links which bind together perceptions that
have been accustomed to occur together.   This fact,
that perceptions and feelings which have frequently
occurred together get linked, so that one calls up the
other, is called the law of Association, and has been
made the basis of scientific Psychology.   According to
these explanations of Locke and Hume (which extended
to the other two groups of rules) all the knowledge we
have that the rules are right, or may be objectively
verified, is really derived from experience; only it is
past experience, which we have had so often and got so
accustomed to that it is now really a part of ourselves.

But Kant, after being staggered for some time by
Hume's explanation, at length said, 'It is impossible
that all your knowledge can have come from experience.

For you know that the axioms of mathematics are absolutely and universally true, and no experience can possibly have told you this. However often you may have found the angles of a triangle amount to two right angles, however accustomed you may have got to this experience, you have no right to know that the angles of every possible triangle are equal to two right angles, nor indeed that those of any one triangle are absolutely and exactly so equal. Now you do know this, and you cannot deny it. You have therefore some knowledge which could not possibly be derived from experience ; it must therefore have come in some other way ; or there is some other source of knowledge besides experience.'

At that time there was no answer whatever to this. For men did think that they knew at least the absolute universality if not the necessity of the mathematical axioms. To any one who admitted the necessity, the argument was even stronger ; for it was clear that no experience could make any approach to supply knowledge of this quality. But if a man felt absolutely sure that two straight lines perpendicular to the same line would never meet, however far produced, he could not maintain against Kant that all knowledge is derived from experience. He was obliged to admit the existence of knowledge à *priori*, that is, knowledge lying ready in the mind from the first, antecedent to all experience.

But now here is a difficulty to be explained, How is it possible that I can have knowledge about objects which is prior to all experience of objects, and which transcends the bounds of possible experience ?

First of all, what do I mean by objects? In the answer to this question lies really Kant's solution of the problem, and I shall endeavour to make this clear by a comparison.

If a man had on a pair of green spectacles, he would see everything green. And if he found out this property of his spectacles, he might say with absolute certainty that while he had those spectacles on everything that he saw without exception would be green.

' Everything that he saw ; ' that is to say, all objects of sight to him. But here it is clear that the word object is relative ; it means a representation that he gets, and has nothing to do with the thing in itself. And the assertion that everything is green would not be an assertion about the things in themselves, but about the representations of them which came to him. The colour of these representations would depend partly on the things outside and partly on his spectacles. It would vary for different things, but there would always be green in it.

Let us modify this example a little. I know for certain that the colour of every object in the universe is made up of colours that lie within the range of the visible spectrum. This is apparently a universal statement, and yet I know it to be true of things which it is impossible that I should ever see. How is this? Why, simply, that my eyes are only affected by light which lies within the range of the visible spectrum. Now I say that this case is only a little modified from the previous one. The green glass lets in a certain range of light ; the range is very little increased when you take it away.   Only in the second case it happens that we are all actually wearing very nearly the same spectacles.   That universal statement which I made is true not only of objects as they appear to me, but also of objects as they appear to you.   It is a statement about objects; that is, about certain representations which we perceive.   It may therefore so far have its origin in the things of which these are representations, or it may have its origin in us.   And we happen to know that in this case it is not a statement about external things, but about our eyes.

Admitting, then, that the objects of our sensations are representations made to us; that their character must therefore be partly dependent upon our own character; what properties of these objects should we naturally suppose to have this origin, to be derived from the constitution of our minds?   Why, clearly, those which are necessary and universal; for only such properties can be so derived, and there is no other way in which they can be known to be universal.

Accordingly, Kant supposes that Space and Time are necessary forms of perception, imposed upon it by the perceiving mind; that things are in space and time as they appear to us, and not in themselves; and that consequently the statement that all things exist in space and time is a statement about the nature of our perception and not about the things perceived.

The word corresponding to experience (*Erfahrung*) is used by Kant nearly in the sense in which I have used *sensation*, to mean the whole phenomenon consisting of the bare message and also of the filling-in, the complete representation which we get of objects.   But it is not

apparently confined to this ; it means not merely the
sensations which I get, but the sensations which I talk
about.   Giving to the word this sense for the present,
we may say that in his theory the form, the general
character, of experience is imposed upon it by two
faculties which we all possess : Intuition and Understanding.   Intuition has necessarily the forms of Space
and Time ; but we are not to say that those properties
of space which are expressed in the geometrical axioms
are all necessitated by the forms of intuition ; for it is
the understanding that supplies us with the pure notions
of quantity, quality, relation, and modality.   It is not
always easy to separate the parts played by these two
faculties in supplying the general rules to which experience conforms ; but it appears, for example, that the
three dimensions of space are given by pure intuition
itself, while the equality of the opposite sides of a parallelogram is only given by help of the understanding.
It is not to our purpose to investigate the difference
between these two faculties, or even to remember that
Kant made a distinction between them.   All that is
important for us is the theory that those general statements upon which the pure sciences are founded,
although really true of objects, that is of representations
made to me, are in fact statements about me and not
about the things in themselves : just as my general
statement about the colours of things was really a statement about my own eyes and not about the things.
And it is just because these statements *are* about me
that I know them to be not only universally, but always
necessarily true about the objects I perceive ; for it is
always the same *me* that perceives them—or at any rate

it is a *me* possessing always the same faculties of representation.

Now observe what it is that this theory does with general statements; what is the means by which it gets rid of them—for it does get rid of them. It makes them into particular statements. Instead of being statements about all possible places and times and things, they are made out to be statements about me, and about other men in so far as they have the same faculties that I have. I want you to notice this transformation particularly, because I shall afterwards endeavour to establish a similar transformation, though in rather a different manner.

In the next place, observe that the question which was proposed by the Critical Philosophy is a perfectly real and important question. It is this:—'Are there any properties of objects in general which are really due to me and to the way in which I perceive them, and which do not belong to the things themselves?' But it seems to me that the method by which Kant attempted to answer this question was not the right method. It consisted in finding what are those characters of experience which we know to be necessary and universal; and concluding that these are characters of me. It requires, therefore, some infallible way of judging what characters are necessary and universal. Now, unfortunately, as I hope to show you, judgments of this kind may very possibly be mistaken If you went up to our man with the green spectacles, and argued with him that since he knew for certain that everything was green, whereas no experience could tell him so, this greenness must be somewhere in the apparatus by which

he perceived things ; there would be just one weakness
in the argument.　He might be mistaken in thinking he
knew that everything was green.　But the proper thing
to do, as it appears to me, would be to take him to a
looking-glass and show him that these spectacles were
actually upon his nose.　And so also in the general
question which is proposed by the Critical Philosophy.
The answer to that question must be sought not in the
subjective method, in the conviction of universality and
necessity, but in the physiological method, in the study
of the physical facts that accompany sensation, and of
the physical properties of the nervous system.　The
materials for this valid criticism of knowledge did not
exist in Kant's time.　I believe that they do exist at
present to such an extent at least as to indicate the
nature of the results which that criticism is to furnish.

The Kantian theory of universal truths was largely,
though not completely, accepted by Whewell, and
applied with considerable detail in his Philosophy of
the Inductive Sciences.　It is necessary to mention him
here, not on account of any important modification that
he introduced into the theory, but because the form into
which he put it has had great influence in directing the
attention of scientific students to the philosophy of
science ; and because by intelligent controversy he con-
tributed very much to the clearing up and development
of an opinion which we have next to consider—that of
Mr. John Stuart Mill.　I can best, I think, set this
opinion before you, if I have permission to quote a short
passage.

‘ To these arguments (of Dr. Whewell, contending

that the axioms could not be known by experience) . . .
a satisfactory answer will, I conceive, be found, if we
advert to one of the characteristic properties of geometrical forms—their capacity of being painted in the imagination with a distinctness equal to reality : in other
words, the exact resemblance of our ideas of form to
the sensations which suggest them.   This, in the first
place, enables us to make (at least with a little practice)
mental pictures of all possible combinations of lines and
angles, which resemble the realities quite as well as any
which we could make on paper ; and in the next place,
make those pictures just as fit subjects of geometrical
experimentation as the realities themselves; inasmuch as
pictures, if sufficiently accurate, exhibit of course all
the properties which would be manifested by the realities at one given instant, and on simple inspection ; and
in geometry we are concerned only with such properties,
and not with that which pictures could not exhibit, the
mutual action of bodies upon one another.   The foundations of geometry would therefore be laid in direct experience, even if the experiments (which in this case
consist merely in attentive contemplation) were practised solely upon what we call our ideas, that is, upon
the diagrams in our minds, and not upon outward
objects.   For in all systems of experimentation we take
some objects to serve as representatives of all which
resemble them ; and in the present case the conditions
which qualify a real object to be the representative of its
class, are completely fulfilled by an object existing only
in our fancy.   Without denying, therefore, the possibility
of satisfying ourselves that two straight lines cannot enclose a space, by merely thinking of straight lines with-

out actually looking at them; I contend that we do not believe this truth on the ground of the imaginary intuition simply, but because we know that the imaginary lines exactly resemble real ones, and that we may conclude from them to real ones with quite as much certainty as we could conclude from one real line to another. The conclusion, therefore, is still an induction from observation. And we should not be authorized to substitute observation of the image in our mind for observation of the reality, if we had not learnt by long-continued experience that the properties of the reality are faithfully represented in the image; just as we should be scientifically warranted in describing an animal which we had never seen from a picture made of it with a daguerreotype; but not until we had learnt by ample experience that observation of such a picture is precisely equivalent to observation of the original.

'These considerations also remove the objection arising from the impossibility of our ocularly following the lines in their prolongation to infinity. For though, in order actually to see that two given lines never meet, it would be necessary to follow them to infinity; yet without doing so we may know that if they ever do meet, or if, after diverging from one another, they begin again to approach, this must take place not at an infinite, but at a finite distance. Supposing, therefore, such to be the case, we can transport ourselves thither in imagination, and can frame a mental image of the appearance which one or both of the lines must present at that point, which we may rely on as being precisely similar to the reality. Now, whether we fix our contemplation upon this imaginary picture, or call to mind the generalizations we have had occasion to make from former ocular observation, we learn by the evidence of experience that a line which, after diverging from another straight line, begins to approach to it, produces the impression on our senses, which we describe by the expression "a bent line," not by the expression "a straight line."'—*Logic*, Book ii., chap. v., s. 5.

Upon this argument I have one very simple remark to make. That 'characteristic property of geometrical forms' is derived from experience ;—we have 'learnt by long-continued experience that the properties of the reality are faithfully represented in the image.' Experience could only tell us this of realities and of images both of which we have experienced. I must know both of two things to know that one faithfully represents the other. Experience then tells me that my mental images of geometrical figures are faithful representations of those realities *which are within the bounds of experience.* But what is to tell me that they are faithful representations of realities that are beyond the bounds of experience? Surely no experience can tell me that.

Again, our notion of *straight* is a combination of several properties, an aggregate of impressions on our senses, which holds together within the limits of experience. But what is to tell us that these impressions hold together beyond the limits of experience?

It seems to me, then, that in admitting the universality of certain statements Mr. Mill knows something which on his own principles he has no right to know.

In the following section Mr. Mill deals with the supposed *necessity* of these truths. Taking this to mean

the inconceivability of the negation of them, he explains
it in somewhat the same way as Hume explained the
idea of cause, namely by means of the law of association.
But that which in Locke and Hume had been merely a
special explanation of particular phenomena has in the
meantime grown into an extensive and most successful
science of Psychology.   It began, as you remember, in
the form of a link between two impressions that occur
frequently together.   Perhaps the most important step
was Hartley's idea of 'mental chemistry;' that the
result of two linked impressions might not put in
evidence either of the components any more than water
exhibits to us the hydrogen and the oxygen which it
contains.   In the hands of James Mill and Mr. Bain
this mode of explanation has been applied with marked
success to a vast number of mental phenomena; so that
when Mr. Mill makes use of it to account for the inconceivability of that which has not yet been experienced,
he is backed by an enormous mass of similar and most
successful explanations.

This view, that the supplementary part of our sensations is an accumulation of past experience, has been
further defended by Mr. Bain in many excellent books.
But there is one respect in which the doctrines of Mr. Mill
and Mr. Bain differ very importantly from the one which
we have next to consider—that of Mr. Herbert Spencer.
He also believes that the whole of our knowledge comes
from experience; but while in the former view this experience is *our own*, and has been acquired during the
lifetime of the individual, in the latter it is not the experience of you or me, but of all our ancestors.   The
perceptions, not only of former generations of men, but

of those lower organisms from which they were originally derived, beginning even with the first molecule that
was complex enough to preserve records of its own
changes; all these have been built into the organism,
have determined its character, and have been handed
down to us by hereditary descent.   The effect of this
upon Kant's doctrine may be best displayed by another
quotation :—

'The universal law that, other things equal, the
cohesion of psychical states is proportionate to the
frequency with which they have followed one another
in experience, supplies an explanation of the so-called
"forms of thought," as soon as it is supplemented by
the law that habitual psychical successions entail some
hereditary tendency to such successions, which, under
persistent conditions, will become cumulative in generation after generation.   We saw that the establishment of
those compound reflex actions called instincts is comprehensible on the principle that inner relations are, by
perpetual repetition, organized into correspondence with
outer relations.   We have now to observe that the
establishment of those consolidated, those indissoluble,
those instinctive mental relations constituting our ideas
of Space and Time, is comprehensible on the same
principle. .   .   .   .

'In the sense, then, that there exist in the nervous
system certain pre-established relations answering to
relations in the environment, there is a truth in the
doctrine of "forms of intuition"—not the truth which
its defenders suppose, but a parallel truth.   Corresponding to absolute external relations, there are established
in the structure of the nervous system absolute internal

relations—relations that are potentially present before birth in the shape of definite nervous connexions; that are antecedent to, and independent of, individual experiences; and that are automatically disclosed along with the first cognitions. And, as here understood, it is not only these fundamental relations which are thus pre-determined; but also hosts of other relations of a more or less constant kind, which are congenitally represented by more or less complete nervous connexions. But these pre-determined internal relations, though independent of the experiences of the individual, are not independent of experiences in general: they have been determined by the experiences of preceding organisms. The corollary here drawn from the general argument is that the human brain is an organised register of infinitely numerous experiences received during the evolution of life, or rather, during the evolution of that series of organisms through which the human organism has been reached. The effects of the most uniform and frequent of these experiences have been successively bequeathed, principal and interest; and have slowly mounted to that high intelligence which lies latent in the brain of the infant—which the infant in after-life exercises and perhaps strengthens or further complicates—and which, with minute additions, it bequeaths to future generations. And thus it happens that the European inherits from twenty to thirty cubic inches more brain than the Papuan. Thus it happens that faculties, as of music, which scarcely exist in some inferior human races, become congenital in superior ones. Thus it happens that out of savages unable to count up to the number of their fingers, and speaking a language containing only nouns and verbs, arise at length our Newtons and our Shakespeares.' —*Principles of Psychology*, § 208, vol. i. pp. 466, 470.

This doctrine of Mr. Spencer's is what I believe to be really the truth about the matter; and I shall have to return to it again by-and-by. But I have a remark to make here. It seems to me that the Kantian dilemma about universal propositions is just as valid now, in spite of these explanations, as it was in his time. How am I to know that the angles of a triangle are exactly equal to two right angles under all possible circumstances; not only in those regions of space where the solar system has been, but everywhere else? The accumulated experience of all my ancestors for a hundred and fifty million years is no more competent to tell me *that* than my own experience of the last five minutes. Either I have some source of knowledge other than experience, and I must admit the existence of *à priori* truths, independent of experience; or I cannot know that any universal statement is true. Now, the doctrine of evolution itself forbids me to admit any transcendental source of knowledge; so that I am driven to conclude in regard to every apparently universal statement, either that it is not really universal, but a particular statement about my nervous system, about my apparatus of thought; or that I do not know that it is true. And to this conclusion, by a detailed examination of various apparently universal statements, I shall in subsequent lectures endeavour to lead you.

II.—KNOWLEDGE AND FEELING.

The following fragment appears to represent what was the conclusion of the series of Lectures as they were delivered in March, 1873. It was found among Professor Clifford's papers without any external indication of its proper context; and as the Lectures now stand after the author's revision, it seems to come in better as an appendix to the first of them. Clifford himself regarded it apparently (note to the Third Lecture in 'Nineteenth Century,' March 1879) as superseded by his article on 'the Nature of Things-in-themselves;' but it contains critical remarks and illustrations which are not there, and it has seemed best to the Editors to let it stand in this place.

IN order to consider at this point what it is that we have arrived at, we must call to mind the point from which we started. We said that the whole of our sensations could not possibly be a message from outside, but that some part at least of them must be a supplement or filling-in of this message, added by ourselves. A theory came before us—that of Mr. Herbert Spencer —according to which this filling-in was accounted for as the product of past experience, which had taken effect on the brains of our ancestors and produced certain changes in them. These changes have gradually moulded the structure of the nervous system which was handed on to us by hereditary descent. There was one obstacle to our acceptance of that theory as a sufficient account of the matter ; namely, that we apparently had some knowledge which could not possibly have been got in that way—knowledge that certain general statements are absolutely and universally true. This obstacle I shall endeavour to remove, by showing that such general statements may be divided into two classes ; of which those in the first class may for all we

know be false, while those in the second class are
general statements only in form, and really are judgments about the apparatus of thought.   If this be so,
we are at liberty to accept the view that all human
knowledge is derived from experience; and that of the
two factors in sensation, that supplement which we
provide of ourselves is a giving out again of what
has originally belonged to the other factor, to experience
proper.   But here a doubt suggests itself which appears
exactly to reverse all that we have done.   We said
there were two factors of experience: that all of it could
not be direct message; and we have come to the conclusion that the two factors are really of the same kind.
But we did not show that any of it *was* direct message
from outside; we only showed that some portions of it
were not.   Suppose it is all supplement, and there is no
message at all!   In that case our two factors will indeed
be reduced to one; but in what sense can we say that
our knowledge is derived from experience?   It will of
course be derived from experience in the large sense,
that is, from sensation; but in the sense in which we
have used the term, as meaning that part of sensation
which is not supplied by ourselves, there will be no
experience for us to derive knowledge from.   This
question then is an extremely important one; for if we
have to admit that there is no real message from without, *all* the sciences will become pure sciences, all knowledge will be *à priori* knowledge; and we may construct
the universe by sitting down and thinking about it.   It
is this question then that I propose to consider for a
short time, a time very much too short for the consideration of it, but perhaps long enough to let me indicate

in some way the kind of answer which is given by an extension of that Physiological Method which we began by using.

We traced the message of sight to the retina of the eye, saying that the only direct message possible is contained in the picture there drawn. But we may go a little further. The picture consists in an aggregate of forms and colours having a certain mode of connexion. It is carried inwards by the optic nerve; but in order to be so carried, it has to undergo a still further transformation. The optic nerve is a great bundle of telegraph wires, each carrying its own message undisturbed by the rest. Each wire only tells what is happening at a particular point of the retina; that is to say, what colour and what intensity the light impinging on the point has. Now in order to tell the colour and intensity, it appears that it must consist of three distinct strands; for it has been made out that every sensation of colour is composed of three simple sensations combined in a certain proportion, this proportion varying from colour to colour. Does then the optic nerve carry the picture itself as a message? It is clear that it cannot; but it may take an account of every point in it, and of their relations of contiguity; that is, it carries an aggregate of elementary messages, which has a point-for-point connexion with the picture, of such a nature as to retain the relations of nextness or contiguity. But the point to notice is that two messages carried by the optic nerve differ only as two chords played upon the same organ, or as two books written in the same alphabet; they are combinations or connected aggregates of the same elementary messages, selected and fastened together in different ways.   The difference is a matter of arrangement and building up ; not a difference of the elements that are built up. This very important step in the theory of sensation was made by Helmholtz, following in the steps of Müller, equally in the case of sight and sound.   It was he who made out clearly that the special nerves of the senses had not absolutely special functions of transmitting their particular sensation as a whole, but that the difference consisted in the various ways of combining together the same elementary nerve-message.   Where, then, are these messages taken?   They are taken to the grey corpuscles within the brain ; and apparently each nerve goes to its own corpuscle, and sets it in commotion with the message. Finally we get this result : that the presence of a picture on the retina involves the commotion of a certain number of grey corpuscles ; the selection of which and the amount of excitement given to each are determined by the picture.   And the same thing happens for every other kind of sensation.   Now the direct knowledge that we get can only be knowledge of this commotion in the grey matter.   For we can tap the telegraph, so to speak, and transmit a false message by it ; and it is found that if the optic nerve be excited either by pressure of the eye or by an electric shock, the sensation of sight is produced, although no light has been present.   The difference, then, of different sensations is made by the difference of the grey corpuscles excited ; and the immediate knowledge that is given to us by experience can only be knowledge of more or less excitement of certain parts of the grey matter.   This applies equally to touch, taste, smell, muscular action, the organic sen-

sations of pain or pleasure. If you and I, then, choose
to contemplate another person, we shall say that the
world which he directly perceives is really inside his
brain, and not outside ; but that corresponding to these
changes that go on in his brain there are certain changes
going on outside of him, and that in many cases there
is such a correspondence of the relations of contiguity in
one case to the relation of contiguity in the other, that
conclusions about the outer world may fairly be drawn
from the world in his brain.

But now, if instead of considering this other person,
I consider myself, the case is rather altered. I shall
conclude by analogy that this world which I directly
perceive is not really outside of me ; that the things
which are apparently made known to me by my perceptions are really themselves only groups of my perceptions ; that the universe which I perceive is made up
of my feelings ; that in fact it is really *me*. And—by
analogy also—I shall conclude that there *is* something
besides this, different from it ; the changes in which
correspond in a certain way to the changes in my
universe. Is it then possible for me to know what that
is ? or is there nothing at all except my feelings ?

If, instead of approaching this question from the
physiological side, we adopt another point of view, it is
not unlikely that we shall be led to the latter conclusion.
If I consider merely my own feelings and ask what
evidence they give of anything beyond them, it seems
to me that I must answer, no evidence at all. This at
least was the answer given by Berkeley in a passage
which has been quoted here before by Professor Huxley,
but will bear quoting again :—

'Some truths there are so near and obvious to the mind that a man need only open his eyes to see them. Such I take this important one to be, viz., that all the choir of heaven and furniture of the earth, in a word all those bodies which compose the mighty frame of the world, have not any subsistence without a mind, that their *being* is to be perceived or known; that consequently so long as they are not actually perceived by me, or do not exist in my mind or that of any other created spirit, they must either have no existence at all, or else subsist in the mind of some Eternal Spirit.' *Principles of Human Knowledge*, § 6.

If I say that such and such things existed at some previous time, I mean that if I had been there I could have perceived them; if I say that there is hydrogen in the sun, I mean that if I could get any of that gas I should be able to burn it in oxygen and produce exactly the same impressions on my senses as those which, in the aggregate, I call *water*.

This doctrine, that the essence of things consists in my perceiving them, is called Idealism. The form of it held by Berkeley, however, is not altogether pure. He believed that no material external world exists; but only spirits exist, thinking beings whose nature consists of conception and volition. Now, from this point of view, fairly accepted, you are only phenomena of my consciousness as much as the rest of the world; I cannot allow the existence of any spirits, but only of one spirit, myself. And even this language is hardly suitable; for why should I give myself a class-name like *spirit*, when I am really the sum-total of the universe? Notwith-

standing this failure to reach complete idealism, the
doctrine of Berkeley, in its positive aspect, is a distinct
and most important step in philosophy; it established
in a security that has never yielded to attack the subjective character of the world of phenomena; that this
world which I perceive *is* my perceptions and nothing
more. Whether there is anything else quite different
which corresponds to it in a certain way, is another
question; Berkeley said there were also spirits.

According to Berkeley, moreover, there exists, besides this world of my perceptions, a particular spirit,
me, that perceives them. To get rid of this imaginary
soul or substance, underlying the succession of my feelings, was the work of Hume. Just as an object, in
Berkeley's theory, is merely a bundle of perceptions
which always occur together, a linked aggregate of
feelings; so, said Hume, out of the swift current of ideas
that succeed one another we construct a unity which we
call Self or Ego. But this, he said, is a pure illusion;
and the ego, when analysed, turns out to be only the
whole complex of my feelings. This, you see, is a step
towards simplification; we had to begin with an external thing which is perceived; then the perception
or feeling; then the soul or self which perceives. With
Berkeley we get rid of the thing perceived; it is
reduced to a bundle of perceptions. With Hume we
get rid also of the perceiving self; it is reduced to
the whole aggregate of feelings, linked together and
succeeding one another in a certain manner.

The step made by Mill is a more complete definition
of the same view, and an explanation by means of the
law of association of the way in which we come to

believe in an external world.   He says that objects are completely described by the phrase, 'permanent possibilities of sensation.'

'The Psychological Theory maintains that there are associations naturally and even necessarily generated by the order of our sensations and of our reminiscences of sensation, which, supposing no intuition of an external world to have existed in consciousness, would inevitably generate the belief, and would cause it to be regarded as an intuition . . . . The conception I form of the world existing at any moment comprises, along with the sensations I am feeling, a countless variety of possibilities of sensation : namely, the whole of those which past observation tells me that I could, under any supposable circumstances, experience at this moment, together with an indefinite and illimitable multitude of others which though I do not know that I could, yet it is possible that I might, experience in circumstances not known to me.   These various possibilities are the important thing to me in the world.   My present sensations are generally of little importance, and are moreover fugitive : the possibilities, on the contrary, are permanent, which is the character that mainly distinguishes our idea of Substance or Matter from our notion of sensation . . . . Matter, then, may be defined, a Permanent Possibility of Sensation.'[1]

In the meanwhile, you observe, the association-theory of the mind had been created ; and it is here applied to defend the position of Hume.   It is worth

[1] J. S. Mill, 'Examination of Sir W. Hamilton's Philosophy,' pp. 192, 193, 198, 2nd edit.

while to notice now where we are. The universe consists of feelings. A certain cable of feelings, linked
together in a particular manner, constitutes me. Similar cables constitute you. That is all there is. But
in the cable of feelings that make up me there are
certain persistent bundles or strands, which occasionally
come to the outside; there are similar strands in the
cables of which you are constituted. These correspond
to external objects; we only think them external for
the reasons assigned.

Now, when we pass to Mr. Herbert Spencer, we
come into the presence of another great department of
science, that has not had so strong an action upon Mr.
Mill; and that is the anatomy of the nervous system.
The effect of investigations in this subject is to analyse
all the various kinds of nervous action into different
combinations of two simple elements; the transmission
of messages along nerve-threads of white matter, and
the excitement of nerve-cells of grey matter Apparently all the nerve-threads are alike, and all the nerve-
cells are alike. The only thing that remains to produce
the very different effects that we observe is the variety
of ways in which selections may be made from the
nerve-cells to be excited at any moment. The direct
effects of nerve-action are the effect on muscular tissue
of contraction or release, and the effect on glands of
secretion.

Here, then, were two great branches of analysis
present to Mr. Spencer: the analysis of mental action
given by the association-theory, which reduced everything to the linking-together of feelings, and the analysis of nervous action supplied by the histologists. It

was his business to supply not merely the link between the two, but an account of their simultaneous evolution. If we find that certain complicated forms of mental action always accompany certain forms of nervous action; if each of these can be reduced into elements, and the relation of each compound to its elements is the same—the bricks different, but the mode of putting them together identical in these two houses—there is a very strong presumption that the element of mental action always accompanies the element of nervous action. But this presumption is converted into knowledge when we have an account of their origin. When the evolution of the living organism is traced upwards from the simplest forms to the most complex, and it is found that the evolution of mind proceeds *pari passu* with it, following the same laws and passing through the same stages, either evolution being expressed as a con tinual building up with the same element, we have actual evidence that the one element goes with the other.

Here, then, is the great advantage of Mr. Herbert Spencer in the study of both orders of facts. He can make any step in analysis of the one help in the analysis of the other. And accordingly he has carried both to an extent which leaves all previous investigators far behind. But you will see at once that we must look at the question of idealism from the physiological point of view. And accordingly he considers that there *is* something dif ferent from our perceptions, the changes in which correspond in a certain way to the changes in the worlds we perceive. He thinks, however, that we can never know what it is; and he says :—

‘We can think of Matter only in terms of Mind.

We can think of Mind only in terms of Matter. When
we have pushed our explorations of the first to the
uttermost limit, we are referred to the second for a
final answer; and when we have got the final answer
of the second, we are referred back to the first for an
interpretation of it. We find the value of x in terms of
y; then we find the value of y in terms of x; and so
on we may continue for ever without coming nearer to
a solution. The antithesis of subject and object, never
to be transcended while consciousness lasts, renders
impossible all knowledge of that Ultimate Reality in
which subject and object are united.'—*Principles of
Psychology*, § 272 (vol. i. p. 627).

Now, the singular character of this realism is that it
is defended from the idealistic point of view, namely,
Mr. Spencer attempts to make my feelings give me
evidence of something which is not included among
them. A careful study of all his arguments to that
effect has only convinced me over again that the attempt
is hopeless. In this respect he differs considerably from
Mr. Shadworth Hodgson, who must be regarded as an
advance, within the British school, in the direction of
Berkeley and Hume. He accepts the analysis of the
individual ego or self into a complex of feeling; and,
like Hume or Mill, makes the universe to consist of
feelings variously bound together. But this is only one
aspect of it and of all contained phenomena. Every
phenomenon has two aspects; in its subjective aspect
it is a feeling, in its objective aspect a quality. But it
is not necessarily a feeling of my consciousness or of
your consciousness; it may be a feeling of the general
or universal consciousness, which is coextensive with

all existence. The universal consciousness bears the same relation to the universal Ego of Schelling or Hegel that the stream of feelings does to the soul; it is an analysis of it into elements.

The important thing here is the conclusion that there is only one world, combined with the analysis of mental phenomena. The German Idealists attempted to construct the world out of very abstract ideas, which are the most complex of all forms of mental action. In this way we did get one world, a mental world ; but the bricks of which it was built were made by the ingenious piling together of houses. I do not think that that process is likely to produce serviceable bricks. Now, Mr. Hodgson's element, feeling, although it seems to imply something too complicated, is yet at least a step in the way of analysis, an indication that analysis is desired.

Can we now. get out of our hobble, and arrive at real knowledge derived from external experience, from messages and not from imagination ? I think we can. But it is necessary to say first what is the character of the knowledge we desire. It will be of the nature of inference, and not of absolute certainty. Now inference depends on the assumption of the uniformity of nature ; and what does this rest on? We cannot infer that which is the ground of all inference ; but although I cannot give you a logical reason for believing it, I can give you a physical explanation of the fact that we all do believe it. We believe a thing when we are prepared to act as if it were true. Now, if you and I had not habitually acted on the assumption of the uniformity of nature from the time when we could act at all, we should not be here to discuss the question. Nature is selecting for survival those individuals and races who act as if she were uniform; and hence the gradual spread of that belief over the civilized world.

This uniformity may be merely a uniformity of *phenomena*, a law relating to my feelings. So long as I only am concerned, it seems to me that the idealist theory is perfectly sufficient. It is quite capable of explaining *me*; but when *you* come into the question, it is utterly at a loss. The distinction between the universal and the individual ego seems to me a merely useless abstraction that throws dust in our eyes. I do believe that you are conscious in the same way as I am; and once that is conceded, the whole idealist theory falls to pieces. For there are feelings which are not my feelings, which are entirely outside my consciousness; so that there is at least an external world. But let us consider now in what way we infer it; why do I believe that there are feelings which are not mine? Because, as I belong to a gregarious race, the greater part of my life consists in acting upon the supposition that it is true.

But now further, have I reason for believing that the changes in this external world correspond in any way with the changes in my world which I perceive? I think so. The complex of feelings which constitutes *you* corresponds in a definite way with the changes which I might perceive in your brain. By inferences that I have previously indicated, I conclude that the ultimate element into which your feeling can be analysed goes with the ultimate element out of which the changes of the nerve-matter in your brain are built up. But physiological action is complicated chemistry, in the same way that chemistry is complicated mechanics. The actions that take place in the brain differ in no way from other material actions, except in their complexity. Conjoin with this the doctrine of Evolution, and you will see evidence that the simplest mental change goes always with the simplest material change, whether in the brain or not. The external world, then, is a complex of mental changes; the ultimate elements into which feeling can be analysed; so simple that the simplest feeling which we can experience is an enormously complex mass of them. Some of these are built up into sufficiently complicated forms to constitute what we call personality, will, consciousness. They all succeed one another according to certain laws; and in virtue of these any conscious aggregate of them is acted upon by the rest; the changes so produced in it are what we call a material world.

There is thus only one world, of elementary feelings; which is perceived by me as my material world. And I am not to look for those complex forms of mental action called intelligence and consciousness, except where I can perceive a correspondingly complex aggregation of matter.

III.—THE POSTULATES OF THE SCIENCE OF SPACE.

In my first lecture I said that, out of the pictures which are all that we can really see, we imagine a world of solid things; and that this world is constructed so as to fulfil a certain code of rules, some called axioms, and some called definitions, and some called postulates, and some assumed in the course of demonstration, but all laid down in one form or another in Euclid's Elements of Geometry. It is this code of rules that we have to consider to-day. I do not, however, propose to take this book that I have mentioned, and to examine one after another the rules as Euclid has laid them down or unconsciously assumed them; notwithstanding that many things might be said in favour of such a course. This book has been for nearly twenty-two centuries the encouragement and guide of that scientific thought which is one thing with the progress of man from a worse to a better state. The encouragement; for it contained a body of knowledge that was really known and could be relied on, and that moreover was growing in extent and application. For even at the time this book was written—shortly after the foundation of the Alexandrian Museum—Mathematic was no longer the merely ideal science of the Platonic school, but had started on her career of conquest over the whole world of Phenomena. The guide; for the aim of every scientific student of every subject was to bring his knowledge of that subject into a form as perfect as that which geometry had attained. Far up on the great mountain of Truth, which all the sciences hope to scale, the foremost of that sacred sisterhood was seen, beckoning to the rest to follow her. And hence she was called, in the dialect of the Pythagoreans, ' the purifier of the reasonable soul.' Being thus in itself at once the inspiration and the aspiration of scientific thought, this Book of Euclid's has had a history as chequered as that of human progress itself. It embodied and systematized the truest results of the search after truth that was made by Greek, Egyptian, and Hindu. It presided for nearly eight centuries over that promise of light and right that was made by the civilized Aryan races on the Mediterranean shores; that promise, whose abeyance for nearly as long an interval is so full of warning and of sadness for ourselves. It went into exile along with the intellectual activity and the goodness of Europe. It was taught, and commented upon, and illustrated, and supplemented, by Arab and Nestorian, in the Universities of Bagdad and of Cordova. From these it was brought back into barbaric Europe by terrified students who dared tell hardly any other thing of what they had learned among the Saracens. Translated from Arabic into Latin, it passed into the schools of Europe, spun out with additional cases for every possible variation of the figure, and bristling with words which had sounded to Greek ears like the babbling of birds in a hedge. At length the Greek text appeared and was translated; and, like other Greek authors, Euclid became an authority. There had not yet arisen in Europe 'that fruitful faculty,' as Mr. Winwood Reade calls it, 'with which kindred spirits contemplate each other's works; which not only takes, but gives; which produces from whatever it receives; which embraces to wrestle, and wrestles to embrace.' Yet it was coming; and though that criticism of first principles which Aristotle and Ptolemy and Galen underwent waited longer in Euclid's case than in theirs, it came for him at last. What Vesalius was to Galen, what Copernicus was to Ptolemy, that was Lobatchewsky to Euclid. There is, indeed, a somewhat instructive parallel between the last two cases. Copernicus and Lobat-

chewsky were both of Slavic origin. Each of them has
brought about a revolution in scientific ideas so great
that it can only be compared with that wrought by the
other. And the reason of the transcendent importance
of these two changes is that they are changes in the
conception of the Cosmos. Before the time of
Copernicus, men knew all about the Universe. They
could tell you in the schools, pat off by heart, all that
it was, and what it had been, and what it would be.
There was the flat earth, with the blue vault of heaven
resting on it like the dome of a cathedral, and the
bright cold stars stuck into it; while the sun and
planets moved in crystal spheres between. Or, among
the better informed, the earth was a globe in the
centre of the universe, heaven a sphere concentric with
it; intermediate machinery as before. At any rate, if
there was anything beyond heaven, it was a void space
that needed no further description. The history of all
this could be traced back to a certain definite time,
when it began; behind that was a changeless eternity
that needed no further history. Its future could be
predicted in general terms as far forward as a certain
epoch, about the precise determination of which there
were, indeed, differences among the learned. But after
that would come again a changeless eternity, which was
fully accounted for and described. But in any case the
Universe was a known thing. Now the enormous effect
of the Copernican system, and of the astronomical
discoveries that have followed it, is that, in place of
this knowledge of a little, which was called knowledge
of the Universe, of Eternity and Immensity, we have
now got knowledge of a great deal more; but we only

call it the knowledge of Here and Now.   We can tell a
great deal about the solar system; but, after all, it is
our house, and not the city.   We can tell something
about the star-system to which our sun belongs; but,
after all, it is our star-system, and not the Universe.
We are talking about Here with the consciousness of a
There beyond it, which we may know some time, but
do not at all know now.   And though the nebular
hypothesis tells us a great deal about the history of
the solar system, and traces it back for a period compared with which the old measure of the duration of
the Universe from beginning to end is not a second to a
century, yet we do not call this the history of eternity.
We may put it all together and call it Now, with the
consciousness of a Then before it, in which things
were happening that may have left records; but we
have not yet read them.   This, then, was the change
effected by Copernicus in the idea of the Universe.   But
there was left another to be made.   For the laws of
space and motion, that we are presently going to
examine, implied an infinite space and an infinite duration, about whose properties as space and time everything was accurately known.   The very constitution of
those parts of it which are at an infinite distance from
us, ' geometry upon the plane at infinity,' is just as well
known, if the Euclidean assumptions are true, as the
geometry of any portion of this room.   In this infinite
and thoroughly well-known space the Universe is
situated during at least some portion of an infinite
and thoroughly well-known time.   So that here we have
real knowledge of something at least that concerns the
Cosmos; something that is true throughout the Im-

mensities and the Eternities. That something Lobatchewsky and his successors have taken away. The
geometer of to-day knows nothing about the nature of
actually existing space at an infinite distance ; he knows
nothing about the properties of this present space in a
past or a future eternity. He knows, indeed, that the
laws assumed by Euclid are true with an accuracy that
no direct experiment can approach, not only in this
place where we are, but in places at a distance from
us that no astronomer has conceived ; but he knows
this as of Here and Now ; beyond his range is a There
and Then of which he knows nothing at present, but may
ultimately come to know more. So, you see, there is a
real parallel between the work of Copernicus and his
successors on the one hand, and the work of Lobatchewsky and his successors on the other. In both of
these the knowledge of Immensity and Eternity is
replaced by knowledge of Here and Now. And in
virtue of these two revolutions the idea of the Universe,
the Macrocosm, the All, as subject of human knowledge,
and therefore of human interest, has fallen to pieces.

It will now, I think, be clear to you why it will not
do to take for our present consideration the postulates of
geometry as Euclid has laid them down. While they
were all certainly true, there might be substituted for
them some other group of equivalent propositions ; and
the choice of the particular set of statements that should
be used as the groundwork of the science was to a
certain extent arbitrary, being only guided by convenience of exposition. But from the moment that the
actual truth of these assumptions becomes doubtful,
they fall of themselves into a necessary order and

classification; for we then begin to see which of them may be true independently of the others. And for the purpose of criticizing the evidence for them, it is essential that this natural order should be taken; for I think you will see presently that any other order would bring hopeless confusion into the discussion.

Space is divided into parts in many ways. If we consider any material thing, space is at once divided into the part where that thing is and the part where it is not. The water in this glass, for example, makes a distinction between the space where it is and the space where it is not. Now, in order to get from one of these to the other you must cross the *surface* of the water; this surface is the boundary of the space where the water is which separates it from the space where it is not. Every *thing*, considered as occupying a portion of space, has a surface which separates the space where it is from the space where it is not. But, again, a surface may be divided into parts in various ways. Part of the surface of this water is against the air, and part is against the glass. If you travel over the surface from one of these parts to the other, you have to cross the *line* which divides them; it is this circular edge where water, air, and glass meet. Every part of a surface is separated from the other parts by a line which bounds it. But now suppose, further, that this glass had been so constructed that the part towards you was blue and the part towards me was white, as it is now. Then this line, dividing two parts of the surface of the water, would itself be divided into two parts; there would be a part where it was against the blue glass, and a part where it was against the white glass. If you travel in thought

along that line, so as to get from one of these two parts
to the other, you have to cross a *point* which separates
them, and is the boundary between them. Every part
of a line is separated from the other parts by points
which bound it. So we may say altogether—

The boundary of a solid (*i.e.*, of a part of space)
is a surface.

The boundary of a part of a surface is a line.

The boundaries of a part of a line are points.

And we are only settling the meanings in which
words are to be used. But here we may make an
observation which is true of all space that we are acquainted with : it is that the process ends here. There
are no parts of a point which are separated from one
another by the next link in the series. This is also
indicated by the reverse process.

For I shall now suppose this point—the last thing
that we got to—to move round the tumbler so as to
trace out the line, or edge, where air, water, and glass
meet. In this way I get a series of points, one after
another ; a series of such a nature that, starting from
any one of them, only two changes are possible that will
keep it within the series : it must go forwards or it
must go backwards, and each of these is perfectly
definite. The line may then be regarded as an aggregate
of points. Now let us imagine, further, a change to take
place in this line, which is nearly a circle. Let us
suppose it to contract towards the centre of the circle,
until it becomes indefinitely small, and disappears. In
so doing it will trace out the upper surface of the water,
the part of the surface where it is in contact with
the air. In this way we shall get a series of circles one

after another—a series of such a·nature that, starting from any one of them, only two changes are possible that will keep it within the series : it must expand or it must contract.   This series, therefore, of circles, is just similar to the series of points that make one circle ; and just as the line is regarded as an aggregate of points, so we may regard this surface as an aggregate of lines. But this surface is also in another sense an aggregate of points, in being an aggregate of aggregates of points. But, starting from a point in the surface, more than two changes are possible that will keep it within the surface, for it may move in any direction.   The surface, then, is an aggregate of points of a different kind from the line.   We speak of the line as a point-aggregate of one dimension, because, starting from one point, there are only two possible directions of change ; so that the line can be traced out in one motion.   In the same way, a surface is a line-aggregate of one dimension, because it can be traced out by one motion of the line ; but it is a point-aggregate of two dimensions, because, in order to build it up of points, we have first to aggregate points into a line, and then lines into a surface.   It requires two motions of a point to trace it out.

Lastly, let us suppose this upper surface of the water to move downwards, remaining always horizontal till it becomes the under surface.   In so doing it will trace out the part of space occupied by the water.   We shall thus get a series of surfaces one after another, precisely analogous to the series of points which make a line, and the series of lines which make a surface.   The piece of solid space is an aggregate of surfaces, and an aggregate of the same kind as the line is of points ; it is

a surface-aggregate of one dimension.   But at the same
time it is a line-aggregate of two dimensions, and a point-
aggregate of three dimensions.   For if you consider a
particular line which has gone to make this solid, a
circle partly contracted and part of the way down, there
are more than two opposite changes which it can undergo.   For it can ascend or descend, or expand or contract,
or do both together in any proportion.   It has just as
great a variety of changes as a point in a surface.   And
the piece of space is called a point-aggregate of three
dimensions, because it takes three distinct motions to get
it from a point.   We must first aggregate points into a
line, then lines into a surface, then surfaces into a solid.

At this step it is clear, again, that the process
must stop in all the space we know of.   For it is not
possible to move that piece of space in such a way as to
change every point in it.   When we moved our line or
our surface, the new line or surface contained no point
whatever that was in the old one ; we started with one
aggregate of points, and by moving it we got an entirely
new aggregate, all the points of which were new.   But
this cannot be done with the solid ; so that the process
is at an end.   We arrive, then, at the result that *space
is of three dimensions.*

Is this, then, one of the postulates of the science of
space?   No ; it is not.   The science of space, as we have
it, deals with relations of distance existing in a certain
space of three dimensions, but it does not at all require
us to assume that no relations of distance are possible
in aggregates of more than three dimensions.   The fact
that there are only three dimensions does regulate the
number of books that we write, and the parts of the

subject that we study : but it is not itself a postulate of the science. We investigate a certain space of three dimensions, on the hypothesis that it has certain elementary properties; and it is the assumptions of these elementary properties that are the real postulates of the science of space. To these I now proceed.

The first of them is concerned with *points*, and with the relation of space to them. We spoke of a line as an aggregate of points. Now there are two kinds of aggregates, which are called respectively continuous and discrete. If you consider this line, the boundary of part of the surface of the water, you will find yourself believing that between any two points of it you can put more points of division, and between any two of these more again, and so on ; and you do not believe there can be any end to the process. We may express that by saying you believe that between any two points of the line there is an infinite number of other points. But now here is an aggregate of marbles, which, regarded as an aggregate, has many characters of resemblance with the aggregate of points. It is a series of marbles, one after another ; and if we take into account the relations of nextness or contiguity which they possess, then there are only two changes possible from one of them as we travel along the series : we must go to the next in front, or to the next behind. But yet it is not true that between any two of them there is an infinite number of other marbles ; between these two, for example, there are only three. There, then, is a distinction at once between the two kinds of aggregates. But there is another, which was pointed out by Aristotle in his Physics and made the basis of a definition of continuity. I have here a

row of two different kinds of marbles, some white and
some black. This aggregate is divided into two parts,
as we formerly supposed the line to be. In the case of
the line the boundary between the two parts is a point
which is the element of which the line is an aggregate.
In this case before us, a marble is the element; but
here we cannot say that the boundary between the two
parts is a marble. The boundary of the white parts is a
white marble, and the boundary of the black parts is a
black marble; these two adjacent parts have different
boundaries. Similarly, if instead of arranging my
marbles in a series, I spread them out on a surface, I
may have this aggregate divided into two portions—a
white portion and a black portion; but the boundary of
the white portion is a row of white marbles, and the
boundary of the black portion is a row of black marbles.
And lastly, if I made a heap of white marbles, and put
black marbles on the top of them, I should have a
discrete aggregate of three dimensions divided into two
parts: the boundary of the white part would be a layer
of white marbles, and the boundary of the black part
would be a layer of black marbles. In all these cases of
discrete aggregates, when they are divided into two
parts, the two adjacent parts have different boundaries.
But if you come to consider an aggregate that you
believe to be continuous, you will see that you think of
two adjacent parts as having the *same* boundary. What
is the boundary between water and air here? Is it
water? No; for there would still have to be a boundary
to divide that water from the air. For the same reason
it cannot be air. I do not want you at present to think
of the actual physical facts by the aid of any molecular

theories; I want you only to think of what appears to be, in order to understand clearly a conception that we all have. Suppose the things actually in contact. If, however much we magnified them, they still appeared to be thoroughly homogeneous, the water filling up a certain space, the air an adjacent space; if this held good indefinitely through all degrees of conceivable magnifying, then we could not say that the surface of the water was a layer of water and the surface of air a layer of air; we should have to say that the same surface was the surface of both of them, and was itself neither one nor the other—that this surface occupied *no* space at all. Accordingly, Aristotle defined the continuous as that of which two adjacent parts have the same boundary; and the discontinuous or discrete as that of which two adjacent parts have direct boundaries.[1]

Now the first postulate of the science of space is that space is a continuous aggregate of points, and not a discrete aggregate. And this postulate—which I shall call the postulate of continuity—is really involved in those three of the six[2] postulates of Euclid for which Robert Simson has retained the name of postulate. You will see, on a little reflection, that a discrete aggregate of points could not be so arranged that any two of

---

[1] Phys. Ausc. V. 3, p. 227, ed. Bekker. Τὸ δὲ συνεχὲς ἔστι μὲν ὅπερ ἐχόμενόν τι, λέγω δ᾽ εἶναι συνεχὲς ὅταν ταὐτὸ γένηται καὶ ἓν τὸ ἑκατέρου πέρας οἷς ἅπτονται, καὶ ὥσπερ σημαίνει τοὔνομα συνέχηται. Τοῦτο δ᾽ οὐχ οἷόν τε δυοῖν ὄντοιν εἶναι τοῖν ἐσχάτοιν.

A little further on he makes the important remark that on the hypothesis of continuity a line is not *made up* of points in the same way that a whole is made up of parts, VI. 1, p. 231. ᾽Αδύνατον ἐξ ἀδιαιρέτων εἶναί τι συνεχές, οἷον γραμμὴν ἐκ στιγμῶν, εἴπερ ἡ γραμμὴ μὲν συνεχές, ἡ στιγμὴ δὲ ἀδιαίρετον.

[2] See De Morgan, in Smith's Dict. of Biography and Mythology, Art. *Euclid*; and in the English Cyclopædia, Art. *Axiom*.

them should be relatively situated to one another in exactly the same manner, so that any two points might be joined by a straight line which should always bear the same definite relation to them. And the same difficulty occurs in regard to the other two postulates. But perhaps the most conclusive way of showing that this postulate is really assumed by Euclid is to adduce the proposition he proves, that every finite straight line may be bisected. Now this could not be the case if it consisted of an odd number of separate points. As the first of the postulates of the science of space, then, we must reckon this postulate of Continuity; according to which two adjacent portions of space, or of a surface, or of a line, have the *same* boundary, viz.—a surface, a line, or a point; and between every two points on a line there is an infinite number of intermediate points.

The next postulate is that of Elementary Flatness. You know that if you get hold of a small piece of a very large circle, it seems to you nearly straight. So, if you were to take any curved line, and magnify it very much, confining your attention to a small piece of it, that piece would seem straighter to you than the curve did before it was magnified. At least, you can easily conceive a curve possessing this property, that the more you magnify it, the straighter it gets. Such a curve would possess the property of elementary flatness. In the same way, if you perceive a portion of the surface of a very large sphere, such as the earth, it appears to you to be flat. If, then, you take a sphere of say a foot diameter, and magnify it more and more, you will find that the more you magnify it the flatter it gets. And you may easily suppose that this process would go on indefinitely;

that the curvature would become less and less the more
the surface was magnified. Any curved surface which
is such that the more you magnify it the flatter it gets,
is said to possess the property of elementary flatness.
But if every succeeding power of our imaginary microscope disclosed new wrinkles and inequalities without
end, then we should say that the surface did not possess
the property of elementary flatness.

But how am I to explain how solid space can have
this property of elementary flatness? Shall I leave it
as a mere analogy, and say that it is the same kind of
property as this of the curve and surface, only in three
dimensions instead of one or two? I think I can get a
little nearer to it than that; at all events I will try.

If we start to go out from a point on a surface,
there is a certain choice of directions in which we may
go. These directions make certain angles with one
another. We may suppose a certain direction to start
with, and then gradually alter that by turning it round
the point: we find thus a single series of directions in
which we may start from the point. According to our
first postulate, it is a continuous series of directions.
Now when I speak of a direction from the point,
I mean a direction of starting; I say nothing about
the subsequent path. Two different paths may have
the same direction at starting; in this case they
will touch at the point; and there is an obvious
difference between two paths which touch and two
paths which meet and form an angle. Here, then, is
an aggregate of directions, and they can be changed
into one another. Moreover, the changes by which
they pass into one another have magnitude, they con-

stitute distance-relations; and the amount of change
necessary to turn one of them into another is called the
angle between them. It is involved in this postulate
that we are considering, that angles can be compared
in respect of magnitude. But this is not all. If we go
on changing a direction of start, it will, after a certain
amount of turning, come round into itself again, and
be the same direction. On every surface which has the
property of elementary flatness, the amount of turning
necessary to take a direction all round into its first
position is the same for all points of the surface. I will
now show you a surface which at one point of it has
not this property. I take this circle of paper from
which a sector has been cut out, and bend it round so
as to join the edges; in this way I form a surface which
is called a *cone*. Now on all points of this surface but
one, the law of elementary flatness holds good. At the
vertex of the cone, however, notwithstanding that there
is an aggregate of directions in which you may start,
such that by continuously changing one of them you
may get it round into its original position, yet the whole
amount of change necessary to effect this is not the
same at the vertex as it is at any other point of the
surface. And this you can see at once when I unroll
it; for only part of the directions in the plane have been
included in the cone. At this point of the cone, then,
it does not possess the property of elementary flatness;
and no amount of magnifying would ever make a cone
seem flat at its vertex.

To apply this to solid space, we must notice that
here also there is a choice of directions in which you
may go out from any point; but it is a much greater

choice than a surface gives you. Whereas in a surface the aggregate of directions is only of one dimension, in solid space it is of two dimensions. But here also there are distance-relations, and the aggregate of directions may be divided into parts which have quantity. For example, the directions which start from the vertex of this cone are divided into those which go inside the cone, and those which go outside the cone. The part of the aggregate which is inside the cone is called a solid angle. Now in those spaces of three dimensions which have the property of elementary flatness, the whole amount of solid angle round one point is equal to the whole amount round another point. Although the space need not be exactly similar to itself in all parts, yet the aggregate of directions round one point is exactly similar to the aggregate of directions round another point, if the space has the property of elementary flatness.

How does Euclid assume this postulate of Elementary Flatness? In his fourth postulate he has expressed it so simply and clearly that you will wonder how anybody could make all this fuss. He says, ' All right angles are equal.'

Why could I not have adopted this at once, and saved a great deal of trouble? Because it assumes the knowledge of a surface possessing the property of elementary flatness in all its points. Unless such a surface is first made out to exist, and the definition of a right angle is restricted to lines drawn upon it—for there is no necessity for the word *straight* in that definition—the postulate in Euclid's form is obviously not true. I can make two lines cross at the vertex of a cone so that the four adjacent angles shall be equal, and yet not one of them equal to a right angle.

I pass on to the third postulate of the science of space—the postulate of Superposition. According to this postulate a body can be moved about in space without altering its size or shape. This seems obvious enough, but it is worth while to examine a little closely into the meaning of it. We must define what we mean by size and by shape. When we say that a body can be moved about without altering its size, we mean that it can be so moved as to keep unaltered the length of all the lines in it. This postulate therefore involves that lines can be compared in respect of magnitude, or that they have a length independent of position; precisely as the former one involved the comparison of angular magnitudes. And when we say that a body can be moved about without altering its shape, we mean that it can be so moved as to keep unaltered all the angles in it. It is not necessary to make mention of the motion of a body, although that is the easiest way of expressing and of conceiving this postulate; but we may, if we like, express it entirely in terms which belong to space, and that we should do in this way. Suppose a figure to have been constructed in some portion of space; say that a triangle has been drawn whose sides are the shortest distances between its angular points. Then if in any other portion of space two points are taken whose shortest distance is equal to a side of the triangle, and at one of them an angle is made equal to one of the angles adjacent to that side, and a line of shortest distance drawn equal to the corresponding side of the original triangle, the distance from the extremity of this to the other of the two points will be equal to the third side of the original triangle, and the two will be equal in all respects; or generally, if a figure has been constructed anywhere, another figure, with all its lines and all its angles equal to the corresponding lines and angles of the first, can be constructed anywhere else. Now this is exactly what is meant by the principle of superposition employed by Euclid to prove the proposition that I have just mentioned. And we may state it again in this short form—All parts of space are exactly alike.

But this postulate carries with it a most important consequence. It enables us to make a pair of most fundamental definitions—those of the plane and of the straight line. In order to explain how these come out of it when it is granted, and how they cannot be made when it is not granted, I must here say something more about the nature of the postulate itself, which might otherwise have been left until we come to criticize it.

We have stated the postulate as referring to solid space. But a similar property may exist in surfaces. Here, for instance, is part of the surface of a sphere. If I draw any figure I like upon this, I can suppose it to be moved about in any way upon the sphere, without alteration of its size or shape. If a figure has been drawn on any part of the surface of a sphere, a figure equal to it in all respects may be drawn on any other part of the surface. Now I say that this property belongs to the surface itself, is a part of its own internal economy, and does not depend in any way upon its relation to space of three dimensions. For I can pull it about and bend it in all manner of ways, so as altogether to alter its relation to solid space ; and yet, if I do not stretch it or tear it, I make no difference whatever in the length of any lines upon it, or in the size of any angles upon it.[1]   I do not in any way alter the figures drawn upon it, or the possibility of drawing figures upon it, *so far as their relations with the surface itself are concerned.*   This property of the surface, then, could be ascertained by people who lived entirely in it, and were absolutely ignorant of a third dimension.   As a point-aggregate of two dimensions, it has in itself properties determining the distance-relations of the points upon it, which are absolutely independent of the existence of any points which are not upon it.

Now-here is a surface which has not that property. You observe that it is not of the same shape all over, and that some parts of it are more curved than other parts.   If you drew a figure upon this surface, and then tried to move it about, you would find that it was impossible to do so without altering the size and shape of the figure.   Some parts of it would have to expand, some to contract, the lengths of the lines could not all be kept the same, the angles would not hit off together. And this property of the surface—that its parts are different from one another—is a property of the surface itself, a part of its internal economy, absolutely independent of any relations it may have with space outside of it.   For, as with the other one, I can pull it about in

---

[1] This figure was made of linen, starched upon a spherical surface, and taken off when dry.   That mentioned in the next paragraph was similarly stretched upon the irregular surface of the head of a bust.   For durability these models should be made of two thicknesses of linen starched together in such a way that the fibres of one bisect the angles between the fibres of the other, and the edge should be bound by a thin slip of paper.   They will then retain their curvature unaltered for a long time.

all sorts of ways, and, so long as I do not stretch it or tear it, I make no alteration in the length of lines drawn upon it or in the size of the angles.

Here, then, is an intrinsic difference between these two surfaces, as surfaces. They are both point-aggregates of two dimensions; but the points in them have certain relations of distance (distance measured always *on* the surface), and these relations of distance are not the same in one case as they are in the other.

The supposed people living in the surface and having no idea of a third dimension might, without suspecting that third dimension at all, make a very accurate determination of the nature of their *locus in quo*. If the people who lived on the surface of the sphere were to measure the angles of a triangle, they would find them to exceed two right angles by a quantity proportional to the area of the triangle. This excess of the angles above two right angles, being divided by the area of the triangle, would be found to give exactly the same quotient at all parts of the sphere. That quotient is called the curvature of the surface; and we say that a sphere is a surface of uniform curvature. But if the people living on this irregular surface were to do the same thing, they would not find quite the same result. The sum of the angles would, indeed, differ from two right angles, but sometimes in excess, and sometimes in defect, according to the part of the surface where they were. And though for small triangles in any one neighbourhood the excess or defect would be nearly proportional to the area of the triangle, yet the quotient obtained by dividing this excess or defect by the area of the triangle would vary from one part of the surface to another. In other words, the curvature of this surface varies from point to point; it is sometimes positive, sometimes negative, sometimes nothing at all.

But now comes the important difference. When I speak of a triangle, what do I suppose the sides of that triangle to be?

If I take two points near enough together upon a surface, and stretch a string between them, that string will take up a certain definite position upon the surface, marking the line of shortest distance from one point to the other. Such a line is called a geodesic line. It is a line determined by the intrinsic properties of the surface, and not by its relations with external space. The line would still be the shortest line, however the surface were pulled about without stretching or tearing. A geodesic line may be *produced*, when a piece of it is given; for we may take one of the points, and, keeping the string stretched, make it go round in a sort of circle until the other end has turned through two right angles. The new position will then be a prolongation of the same geodesic line.

In speaking of a triangle, then, I meant a triangle whose sides are geodesic lines. But in the case of a spherical surface—or, more generally, of a surface of constant curvature—these geodesic lines have another and most important property. They are *straight*, so far as the surface is concerned. On this surface a figure may be moved about without altering its size or shape. It is possible, therefore, to draw a line which shall be of the same shape all along and on both sides. That is to say, if you take a piece of the surface on one side of such a line, you may slide it all along the line and it will fit; and you may turn it round and apply it to the other side, and it will fit there also. This is Leibnitz's definition of a straight line, and, you see, it has no meaning except in the case of a surface of constant curvature, a surface all parts of which are alike.

Now let us consider the corresponding things in solid space. In this also we may have geodesic lines; namely, lines formed by stretching a string between two points. But we may also have geodesic surfaces; and they are produced in this manner. Suppose we have a point on a surface, and this surface possesses the property of elementary flatness. Then among all the directions of starting from the point, there are some which start *in the surface*, and do not make an angle with it. Let all these be prolonged into geodesics; then we may imagine one of these geodesics to travel round and coincide with all the others in turn. In so doing it will trace out a surface which is called a geodesic surface. Now in the particular case where a space of three dimensions has the property of superposition, or is all over alike, these geodesic surfaces are *planes*. That is to say, since the space is all over alike, these surfaces are also of the same shape all over and on both sides; which is Leibnitz's definition of a plane. If you take a piece of space on one side of such a plane, partly bounded by the plane, you may slide it all over the plane, and it will fit; and you may turn it round and apply it to the other side, and it will fit there also. Now it is clear that this definition will have no meaning unless the third postulate be granted. So we may say that when the postulate of Superposition is true, then there are planes and straight lines; and they are defined as being of the same shape throughout and on both sides.

It is found that the whole geometry of a space of three dimensions is known when we know the curvature of three geodesic surfaces at every point. The third postulate requires that the curvature of all geodesic surfaces should be everywhere equal to the same quantity.

I pass to the fourth postulate, which I call the postulate of Similarity. According to this postulate, any figure may be magnified or diminished in any degree without altering its shape. If any figure has been constructed in one part of space, it may be reconstructed to any scale whatever in any other part of space, so that no one of the angles shall be altered, though all the lengths of lines will of course be altered. This seems to be a sufficiently obvious induction from experience; for we have all frequently seen different sizes of the same shape; and it has the advantage of embodying the fifth and sixth of Euclid's postulates in a single principle, which bears a great resemblance in form to that of Superposition, and may be used in the same manner. It is easy to show that it involves the two postulates of Euclid: 'Two straight lines cannot enclose a space,' and 'Lines in one plane which never meet make equal angles with every other line.'

This fourth postulate is equivalent to the assumption that the constant curvature of the geodesic surfaces is zero; or the third and fourth may be put together, and we shall then say that the three curvatures of space are all of them zero at every point.

The supposition made by Lobatchewsky was, that the three first postulates were true, but not the fourth. Of the two Euclidean postulates included in this, he admitted one, viz., that two straight lines cannot enclose a space, or that two lines which once diverge go on diverging for ever. But he left out the postulate about parallels, which may be stated in this form. If through a point outside of a straight line there be drawn another, indefinitely produced both ways; and if we turn this second one round so as to make the point of intersection travel along the first line, then at the very instant that this point of intersection disappears at one end it will reappear at the other, and there is only one position in which the lines do not intersect. Lobatchewsky supposed, instead, that there was a finite angle through which the second line must be turned after the point of intersection had disappeared at one end, before it reappeared at the other. For all positions of the second line within this angle there is then no intersection. In the two limiting positions, when the lines have just done meeting at one end, and when they are just going to meet at the other, they are called parallel; so that two lines can be drawn through a fixed point parallel to a given straight line. The angle between these two depends in a certain way upon the distance of the point from the line. The sum of the angles of a triangle is less than two right angles by a quantity proportional to the area of the triangle. The whole of this geometry is worked out in the style of Euclid, and the most interesting conclusions are arrived at; particularly in the theory of solid space, in which a surface turns up which is not plane relatively to that space, but which, for purposes of drawing figures upon it, is identical with the Euclidean plane.

It was Riemann, however, who first accomplished the task of analysing all the assumptions of geometry, and showing which of them were independent. This very disentangling and separation of them is sufficient to deprive them for the geometer of their exactness and necessity; for the process by which it is effected consists in showing the possibility of conceiving these suppositions one by one to be untrue; whereby it is clearly made out how much is supposed. But it may be worth while to state formally the case for and against them.

When it is maintained that we know these postulates to be universally true, in virtue of certain deliverances of our consciousness, it is implied that these deliverances could not exist, except upon the supposition that the postulates are true. If it can be shown, then, from experience that our consciousness would tell us exactly the same things if the postulates are not true, the ground of their validity will be taken away. But this is a very easy thing to show.

That same faculty which tells you that space is continuous tells you that this water is continuous, and that the motion perceived in a wheel of life is continuous. Now we happen to know that if we could magnify this water as much again as the best microscopes can magnify it, we should perceive its granular structure. And what happens in a wheel of life is discovered by stopping the machine. Even apart, then, from our knowledge of the way nerves act in carrying messages, it appears that we have no means of knowing anything more about an aggregate than that it is too fine-grained for us to perceive its discontinuity, if it has any.

Nor can we, in general, receive a conception as positive knowledge which is itself founded merely upon inaction. For the conception of a continuous thing is of that which looks just the same however much you magnify it. We may conceive the magnifying to go on to a certain extent without change, and then, as it were, leave it going on, without taking the trouble to doubt about the changes that may ensue.

In regard to the second postulate, we have merely to point to the example of polished surfaces. The smoothest surface that can be made is the one most completely covered with the minutest ruts and furrows. Yet geometrical constructions can be made with extreme accuracy upon such a surface, on the supposition that it is an exact plane. If, therefore, the sharp points, edges, and furrows of space are only small enough, there will be nothing to hinder our conviction of its elementary flatness. It has even been remarked by Riemann that we must not shrink from this supposition if it is found useful in explaining physical phenomena.

The first two postulates may therefore be doubted on the side of the very small. We may put the third and fourth together, and doubt them on the side of the very great. For if the property of elementary flatness exist on the average, the deviations from it being, as we have supposed, too small to be perceived, then, whatever were the true nature of space, we should have exactly the conceptions of it which we now have, if only the regions we can get at were small in comparison with the areas of curvature. If we suppose the curvature to vary in an irregular manner, the effect of it might be very considerable in a triangle formed by the nearest fixed stars; but if we suppose it approximately uniform to the limit of telescopic reach, it will be restricted to very much narrower limits. I cannot perhaps do better than conclude by describing to you as well as I can what is the nature of things on the supposition that the curvature of all space is nearly uniform and positive.

In this case the Universe, as known, becomes again a valid conception; for the extent of space is a finite number of cubic miles.[1] And this comes about in a curious way. If you were to start in any direction whatever, and move in that direction in a perfect straight line according to the definition of Leibnitz; after travelling a most prodigious distance, to which the parallactic unit —200,000 times the diameter of the earth's orbit— would be only a few steps, you would arrive at—this place. Only, if you had started upwards, you would appear from below. Now, one of two things would be true. Either, when you had got half-way on your journey, you came to a place that is opposite to this, and which you must have gone through, whatever direction you started in; or else all paths you could have taken diverge entirely from each other till they meet again at this place. In the former case, every two straight lines in a plane meet in two points, in the

---

[1] The assumptions here made about the *Zusammenhang* of space are the simplest ones, but even the finite extent does not follow necessarily from uniform positive curvature; as Riemann seems to have supposed.

latter they meet only in one. Upon this supposition of a positive curvature, the whole of geometry is far more complete and interesting; the principle of duality, instead of half breaking down over metric relations, applies to all propositions without exception. In fact, I do not mind confessing that I personally have often found relief from the dreary infinities of homaloidal space in the consoling hope that, after all, this other may be the true state of things.

IV.—THE UNIVERSAL STATEMENTS OF ARITHMETIC.

WE have now to consider a series of alleged universal statements, the truth of which nobody has ever doubted. They are statements belonging to arithmetic, to the science of quantity, to pure logic, and to a branch of the science of space which is of quite recent origin, which applies to other objects besides space, and is called the analysis of position. I shall endeavour to show that the case of these statements is entirely different from that of the statements about space which I examined in my last lecture. There were four of those statements: that the space of three dimensions which we perceive is a continuous aggregate of points, that it is flat in its smallest parts, that figures may be moved in it without alteration of size or shape, and that similar figures of different sizes may be constructed in it. And the conclusion which I endeavoured to establish about these statements was that, for all we know, any or all of them may be false. In regard to the statements we have now to examine, I shall not maintain a similar doctrine; I shall only maintain that, for all we know, there may be times and places where they are unmeaning and inapplicable. If I am asked what two and two make, I shall not reply that it depends upon circumstances, and that they make sometimes three and sometimes five; but I shall endeavour to show that unless our experience had certain definite characters, there would be no such conception as two, or three, or four, and still less such a conception as the adding together of two numbers; and that we have no warrant for the absolute universality of these definite characters of experience.

In the first place it is clear that the moment we use language at all, we may make statements which are apparently universal, but which really only assign the meaning of words. Whenever we have called a thing by two names, so that every individual of a certain class bears the name A and also the name B, then we may affirm the apparently universal proposition that every A is B. But it is really only the particular proposition that the name A has been conventionally settled to have the same meaning as the name B. I may, for example, enunciate the proposition that all depth is profundity, and all profundity is depth. This statement appears to be of universal generality; and nobody doubts that it is true. But for all that it is not a statement of some fact which is true of nature as a whole; it is only a statement about the use of certain words in the English language. In this case the meaning of the two words is co-extensive; one means exactly as much as, and no more than, the other. But if we suppose the word *crow* to mean a black bird having certain peculiarities of structure, the statement,

'All crows are black,' is in a similar case. For the word *black* has part of the meaning of the word *crow*; and the proposition only states this connexion between the two words. Are the propositions of arithmetic, then, mere statements about the meanings of words? No; but these examples will help us to understand them. Language is part of the apparatus of thought; it is that by which I am able to talk to myself. But it is not all of the apparatus of thought; and just as these apparently general propositions, 'All crows are black,' 'All depth is profundity,' are really statements about language, so I shall endeavour to show that the statements of arithmetic are really statements about certain other apparatus of thought.

We know that six and three are nine. Wherever we find six things, if we put three things to them, there are nine things altogether. The terms are so simple and so familiar, that it seems as if there were no more to be said, as if we could not examine into the nature of these statements any further.

No more there is, if we are obliged to take words as they stand, with the complex meanings which at present belong to them. But the real fact is that the meanings of *six* and *three* are already complex meanings, and are capable of being resolved into their elements. This resolution is due—I believe equally and independently —to two great living mathematicians, by whose other achievements this country has retained the scientific position which Newton won for her at a time of fierce competition when no ordinary genius could possibly have attained it. The conception of *number*, as represented by that word and also by the particular signs,

three, six, and so on, has been shown to embody in
itself a certain proposition, upon the repetition of which
the whole science of arithmetic is based.   By means
of this remark of Cayley and Sylvester, we are able to
assign the true nature of arithmetical propositions, and
to pass from thence by an obvious analogy to those
other cases that we have to consider.

What do I do to find out that a certain set of things
are six in number?   I count them ; and all counting,
like the names of numbers, belongs first to the fingers.
Now this is the operation of counting ; I take my
fingers in a certain definite order—say I begin with
the thumb of each hand, and with the right hand.
Then I lay my fingers in this order upon the things to
be counted ; or if they are too far away, I imagine that
I lay them.   And I observe what finger it is that is
laid upon the *last* thing, and call the things by the
name of this finger.   In the present case it is the thumb
of my left hand ; and if we were savages, that thumb
would be called six.   At any rate, if the order of my
fingers is settled beforehand, and known to everybody,
I can quite easily make the statement, 'Here are six
things,' by holding up the thumb of my left hand.

But, if I have only gone through this process once,
there is already a great assumption made.   For, although
the order in which I use my fingers is fixed, there is
nothing at all said about the order in which the things
are touched by them.   It is assumed that if the things
are taken in any other order and applied to my fingers,
the last one so touched will be the thumb of my left
hand.   If this were not true, or were not assumed, the
word 'number' could not have its meaning.   There is

implied and bound up in that word the assumption that
a group of things comes ultimately to the same finger
in whatever order they are counted. This is the proposition of which I spoke as the foundation of the whole
science of number. It is involved not only in the
general term ' number,' but also in all the particular
names of numbers; and not only in these words, but in
the sign of holding up a finger to indicate how many
things there are.

Let us now look in this light at the statement that
six and three are nine. I have counted a group of
things, and come to the conclusion that there are six of
them. I have already said, therefore, that they may
be counted in any order whatever and will come to the
same number, six. I have counted another distinct
group, and come to the conclusion that there are three
of them. Then I put them all together and count them.
Now, without seeing or knowing any more of the things
than is implied in the previous statements, I can already
count them *in a certain order* with my fingers. For I
will first suppose the six to be counted; the last of
them, by hypothesis, is attached in thought to the thumb
of my left hand. Now I will count the other three;
they are then attached, by hypothesis, to the first three
fingers of my right hand. I can now go on counting
the aggregate group by attaching to these three fingers
the successive fingers of my left hand; for thus I shall
attach the remaining three things to those fingers. I
find in this way that the last of them comes to the fourth
finger of my left hand, counting the thumb as first; and
I know therefore that if the aggregate group has any
number at all, that number must be *nine*.

But this is an operation performed on my fingers; and the statement that we have founded on it must therefore be, at least in part, a statement about my counting apparatus. We may easily understand what is meant by saying that six and three are nine *on my fingers*, independently of any other things than those; this is a particular statement only. The statement we want to examine is that this is equally true of any two distinct groups whatever of six things and three things, which appears to be a universal statement. Now I say that this latter statement can be resolved into two as follows:—

1. The particular statement aforesaid : six and three are nine *on my fingers*.

2. If there is a group of things which can be attached to certain of my fingers, one to each, and another group of things which can be attached to certain other of my fingers, one to each, then the compound group can be attached to the whole set of my fingers that have been used, one to each.

Now this latter, it seems to me, is a tautology or identical proposition, depending merely upon the properties of language. The arithmetical proposition, then, is resolved or analysed in this way into two parts—a particular statement about my counting apparatus, and a particular statement about language ; and it is not really general at all. But this, it is important to notice, is not the complete solution of the problem ; there is a certain part of it reserved. For I only arrive at the number nine by certain definite ways of counting ; I must count the six things first and then the three things after them. And I only arrive at the result that if the aggregate group of things has any number at all, that number is nine. It is not yet proved that they may be counted in any order whatever, and will always come to that number. Here, then, we are driven back to consider the nature of that fundamental assumption that the number of any finite group of distinct things is independent of the order of counting. Here is a proposition apparently still more general than any statement about the sum of two numbers. Do I or do I not know that this is true of very large numbers? Consider, for example, the molecules of water in this glass. According to Sir William Thomson, if a drop of water were magnified to the size of the earth it would appear coarser-grained than a heap of small shot, and finer-grained than a heap of cricket-balls. We may therefore soon find that the number of molecules in this glass very far transcends our powers of conception. Do I know that if these molecules were counted in a certain order, and then counted over again in a certain other order, the results of these two countings would be the same? For the operations are absolutely impossible in anybody's lifetime. Can I know anything about the equivalence of two impossible operations, neither of which can be conceived except in a symbolic way? And if I do, how is it possible for this knowledge to come from experience?

I reply that I do know it; that such knowledge of things as there is in it has come from experience; and that, in fact, it is made up of a particular statement and a conventional use of words. These views will appear paradoxical; but the justification of them is to be found in the analysis of that fundamental assumption which lies at the basis of the idea of number.

In the first place I shall *prove* this fundamental assumption in the case of the number six—that is to say, I shall show that it is involved in suppositions which are already made before there is any question of it. The proposition we have to prove is: if a group of distinct things comes to six when counted in a certain order, it will come to six when counted in any other order. I say that the proposition is involved in the meaning of the phrase *distinct things*, and may be got out of it by help of a particular observation.

What, then, is meant by ' a group of distinct things'? That they are all distinct from one another, or that any one and any other of them make two. That is, if they are attached to two of my fingers in a certain order, they can also be attached to the same two fingers in the other order. Now, for simplicity, let us take the letters in the word *spring*, and count them first as they occur in that word and then in the alphabetical order. I say that, merely on the supposition that they are distinct from one another, I can change one order into the other while I use the same fingers to attach them to.

1	2	3	4	5	6
S	P	R	I	N	G
G	P	R	I	N	S
G	I	R	P	N	S
G	I	N	P	R	S

In the new order I want G to be first; now the letters G and S are by hypothesis distinct, they are two letters. I can therefore interchange the fingers to which they are attached without using more or fewer fingers than before. The same thing is true by hypothesis of I and P, and finally of N and R. By these steps, then, I have changed one order into the other without altering the fingers used in counting—that is, without altering the number. And each of these steps is involved in the meaning of the words *distinct things*—that is, it is made possible by the assumptions which these words involve. But now observe further: how do I know that I can make enough steps to effect the whole change required? In this way. It is given to me in the hypothesis that the things have been counted once; I can therefore go to them one by one till I come to the end. But as I go to each one I can substitute by this process the new one which is wanted in its stead in such a way that the required new order shall hold good behind me. Thus you see that all the steps are involved in the word *distinct*, by the help of an observation on two of my fingers ; and that the possibility of a sufficient number of them to effect the change is involved in the hypothesis that the things have been once counted. Here I have two distinct statements : the first is that the things are distinct, and have been once counted as six ; the second is that in another order they come to the same. When I examine into the meaning of these, I find that they are not statements of different facts, but different statements of the same facts. That one statement is true, or that the other statement is true,—that is a matter of experience ; but that if one is true the other is true, that is a matter of language.

I have only spoken, however, of the particular number six ; how am I to extend these remarks to numbers which cannot be counted, like the number of molecules in this glass of water? In the first place we all know that cultivated races do not count directly with their fingers, but with the names of them—with the words one, two, three, four. Next, this system of names has been extended indefinitely, by a process to which no end can be conceived. But the remarks that we have made about finger-counting will hold good in every case in which the actual counting can be performed. Now in those cases in which this is not true—in the case of a billion, for example—we have two statements made, neither of which can be adequately represented in thought, but which, in so far as they can be represented, are identical statements. That there are a billion grains of sand in a certain heap, provided they be counted in a certain order—this is a supposition which can only be made symbolically. But in so far as it can be made, it is the same supposition as that they also come to a billion in any other order. Any step towards the representation in thought of the one statement is the same step towards the representation in thought of the other; and I do not know any other way in which two symbolic statements can be statements of the same facts. Pure water is the same thing as *aqua pura*; and wherever there are seventy thousand million tons of pure water there are seventy thousand million tons of *aqua pura*. I know that to be true, but it is not a statement of fact: it is a statement about language, notwithstanding that the language is used to symbolize that which cannot be actually represented in thought. So when I say of these molecules of water, 'If they are distinct things, the number of them counted in one order is equal to the number of them counted in any other order,' I make a supposition which I cannot realize in thought. I cannot possibly call up those molecules two and two to observe their distinctness. The supposition is only represented symbolically by language ; but the statement that follows it is the same supposition represented symbolically by other language ; and the equivalence of the two is, after all, a statement about language and not about facts.

But you will say, I do know that these molecules are distinct things ; and so I am able to make these equivalent statements about them. I know that they have a definite number, which is the same however they are counted.

Yes, I know that they are distinct things ; but only by inference, on the assumption of the uniformity of nature ; and about that there is more to be said. The distinctness of things—the fact that one thing and one thing make two—this belongs to our experience. It is a fact that impressions hang together in groups which persist as groups, and in virtue of this persistence we call them things. So long as our experience consists of things, we may build out of it the conceptions of number ; and the nature and connexion of these conceptions are determined by the primary sensation of things as individuals. Now there can, I think, be no doubt that the experience of a hundred or a hundred and fifty million years has so modified our nervous systems that without total disruption of them we cannot cease to aggregate our perceptions into more or less persistent groups ; the continuity of things has become a form of sense. If we were placed in circumstances where these aggregations of feeling were not naturally produced, where perceptible things were not continuous in their changes, we should go on perceiving chaos as made of individual things for at least some time.   But the perception would be a false one, and in acting upon it we should come to grief. Meanwhile, however, the science of number would be perfectly true of our perceptions, though practically inapplicable to the world.

To sum up, then, we carry about with us a certain apparatus of counting, which was primarily our fingers, but is now extended into a series of signs which we can remember in a certain order—the names of numbers. Our language is so formed as to make us able to talk to ourselves about the results of counting.   The propositions of arithmetic are compounded in general of two parts ; a statement about the counting apparatus, and a statement about the different ways of describing its results.

But before quite leaving this let us fix our attention for a short time on the mode of use of the counting apparatus.   The operation of counting a certain group of things consists in assigning one of these numeral words to each of them ; in establishing a correspondence between two groups, so that to every thing or element of the one group is assigned one particular thing or element of the other.   There is here a one-to-one correspondence of two aggregates, one of which is carried about as a standard ; and the propositions arrived at are always of this kind :—if a group of things can have this correspondence with the standard group, then those properties of the standard group which are carried over by the correspondence will belong to the new group.   Now this establishment of correspondence between two aggregates and investigation of the properties that are carried over by the correspondence may be called the central idea of modern mathematics : it runs through the whole of the pure science and of its applications. It may be conceived, therefore, that propositions which are apparently as general and certain as those we have discussed to-day may be analysed in the same manner, and shown to be really statements about the apparatus of thought.

In my second lecture I endeavoured to explain the difference between a discrete and a continuous aggregate. In a row of marbles, which is a discrete aggregate, we can find between any two marbles only a finite number of others, and sometimes none at all. But if two points are taken on a line, the hypothesis of continuity supposes that there is no end to the number of intermediate points that we can find. Precisely the same difference holds good between *number* and *continuous quantity*. The several marbles, beginning at any one of them, may be numbered one, two, three, &c. ; and the number attached to each marble will be the number of marbles from the starting-point to that marble inclusive. If the points on a line are regarded as forming a continuous aggregate, then lengths measured along the line from an arbitrary point on it are called *continuous quantities*. So also, if the instants of time are regarded as forming a continuous aggregate—that is, if we suppose that between any two instants there is no end to the number of intermediate ones that might be found —then intervals or lengths of time will be continuous quantities. And just as we may attach our numbers one by one to the marbles which form a discrete aggregate, so we may attach continuous quantities (or shortly *quantities*) one by one to the points which form a continuous aggregate. Thus to the point P will be attached the quantity or length A P. And we see thus that between

any two quantities there may be found an infinite number of intermediate quantities, while between two numbers there can only be found a finite number of intermediate numbers, and sometimes none at all. That is to say, continuous quantities form a continuous aggregate, while numbers form a discrete aggregate. Thus the science of quantity is a totally different thing from the science of number.

Notwithstanding that this difference was clearly perceived by the ancients, attempts have constantly been made by the moderns to treat the two sciences as one, and to found the science of quantity upon the science of number. The method is to treat rational fractions as a necessary extension of numerical division, and then to deal with incommensurable quantities by way of continual approximation. In the science of number, while five-sevenths of fourteen has a meaning, namely, ten, five-sevenths of twelve is nonsense. Let us then treat it as if it were sense, and see what comes of it. A repetition of this process with every impossible operation that occurs is supposed to lead in time to continuous quantities. The results of such attempts are the substitution of algebra for the fifth book of Euclid or some equivalent doctrine of continuous ratios, and the substitution of the differential calculus for the method of fluxions. For my own part, I believe this method to be logically false and educationally mischievous. For reasons too long to give here, I do not believe that the provisional use of unmeaning arithmetical symbols can ever lead to the science of quantity; and I feel sure that the attempt to found it on such abstractions obscures its true physical nature. The science of number is founded on the hypothesis of the distinctness of things; the science of quantity is founded on the totally different hypothesis of continuity. Nevertheless, the relations between the two sciences are very close and extensive. The scale of numbers is used, as we shall see, in forming the mental apparatus of the scale of quantities, and the fundamental conception of equality of ratios is so defined that it can be reasoned about in the terms of arithmetic.[1] The operations of addition and subtraction of quantities are closely analogous to the operations of the same name performed on numbers and follow the same laws. The composition of ratios includes numerical multiplication as a particular case, and combines in the same way with addition and subtraction. So close and far-reaching is this analogy that the processes and results of the two sciences are expressed in the same language, verbal and symbolical, while no confusion is produced by this ambiguity of meaning, except in the minds of those who try to make familiarity with language do duty for knowledge of things.

Just as in operations of counting there is a com-

[1] Defining a fraction as the ratio of two numbers, Euclid's definition of proportion is equivalent to the following:—Two quantity-ratios are equal if every fraction is either less than both, equal to both, or greater than both of them.

parison of some aggregate of discrete things with a scale of numbers carried about with us as a standard, so in operations of measuring, real or ideal, there is comparison of some piece of a continuous thing with a scale of quantities. We may best understand this scale by the example of time. To indicate exactly the time elapsed from the beginning of the century to some particular instant of to-day, it is necessary and sufficient to name the date and point to the hands of a clock which was going right and was stopped at that instant. This is equivalent to saying that the whole quantity of time consists, first, of a certain number of hours, specified by comparison with the scale of numbers already constructed, and, secondly, of a certain part of an hour, which being a continuous quantity can only be adequately specified by another continuous quantity representing it on some definite scale. In the present case this is conveniently taken to be the arc of a circle described by the point of the minute-hand. On the scale in which that whole circumference represents an hour, this arc represents the portion of an hour which remains to be added. With the help of the scale of numbers, then, any assigned continuous quantity will serve as a standard by which the whole scale of quantities may be represented. And when we assert that any theorem, *e.g.*, the binomial theorem, is true of all quantities whatever, whether of length, of time, of weight, or of intensity, we really assert two things : first, this theorem is true *on the standard* ; secondly, relations of the measures of quantities on the standard are relations of the quantities themselves. The first is (in regard to the *kind* of quantity) a particular statement ;

the second is involved in the meaning of the words 'quantity' and 'measurement.'

But the most familiar and perhaps the most natural form of the scale of quantities is that in which it is supposed to be marked off on a straight line, starting from an arbitrarily assumed point which is called the *origin.* If we make the four assumptions of Euclidean or parabolic geometry, the position of every point in space may be specified by three quantities marked off on three straight lines at right angles to each other, their common point of intersection being taken as origin, and the direction in which each of the quantities is measured being also assigned. Namely, these three quantities are the distances from the origin to the feet of perpendiculars let fall from the point to be specified on the three straight lines respectively. In all space of three dimensions the position of a point may be specified in general by a set of three quantities; but two or more points may belong to the same set of quantities, or two or more sets may specify the same point; and there may be exceptional sets specifying not one point, but all the points on a curve or surface, and exceptional points belonging to an infinite number of sets of quantities subject to some condition. There are three kinds of space of three dimensions in which this specification is *unique,* one point for one set of quantities, one set of quantities for every point, and *without any exceptional cases.* These three are the hypothetical space of Euclid, with no curvature; the space of Lobatchewsky, with constant negative curvature; and the space I described at the end of my second lecture, with constant positive curvature. In only one of these, the space of Euclid, are the three quantities specifying a point actual distances of the point from three planes. In this alone we have a simple and direct representation of the scale of quantities. Now, if we remember that the scale of quantities is a mental apparatus depending only on the first of our four assumptions about space, we may see in this distinctive property of Euclidean space a probable origin for the curious opinion that it has some *à priori* probability or even certainty, as the true character of the universe we inhabit, over and above the observation that within the limits of experience that universe does approximately conform to its rules. It has even been maintained that if our space has curvature, it must be contained in a space of more dimensions and no curvature. I can think of no grounds for such an opinion except the property of flat spaces which I have just mentioned.

END OF THE FIRST VOLUME.

LONDON : PRINTED BY
SPOTTISWOODE AND CO., NEW-STREET SQUARE
AND PARLIAMENT STREET

Printed by Printforce, United Kingdom